THE GREAT BRAIN ROBBERY

By

DAVID C. C. WATSON

MOODY PRESS
CHICAGO

Printed in the United States of America

To

the late Mrs. Cicely Vaughan Wilkes
of St. Cyprian's School, Eastbourne,
who first taught me to read, mark, learn, relish,
and inwardly digest the Bible.

The School Motto
Forsan et haec olim meminisse juvabit.
"Perhaps one day it will be pleasant to remember these
things too."

ACKNOWLEDGMENTS

The gestation period of this book has been more than a decade, and my thanks are due to many friends who over the years have lent encouragement, advice, and suggestions. In particular I should mention Professor Enoch of Madras and Dr. T. N. Sterrett of Bangalore; and, more recently, Dr. D. B. Gower of Guy's Hospital, Dr. Monty White of Cardiff, and Dr. Arthur Jones of Birmingham, England. For the views here expressed, of course, and for any errors of fact or judgment, I alone am responsible.

To the Reverend P. M. Masters I am indebted for the publication of some of these chapters in the *Evangelical Times*, under his editorship.

Dr. Morris and Professor Whitcomb have kindly given me permission to quote extensively from their books. I believe that one day the whole Church will recognize the debt we owe them for their immense industry, zeal, and wisdom in the cause of Christ.

CONTENTS

INTRODUCTION

Crimes should be exposed not when they are being talked about but while they are being committed.

ALEXANDER SOLZHENITSYN

This is a people robbed and spoiled.

ISAIAH 42:22

The Guinness Book of Records is a fascinating compilation. We are not surprised to find that the train robbery near Cheddington, England, on August 8, 1963 was the greatest ever. And the next page informs us of the greatest forgery: £150 million worth of £5 notes were printed by the German government in 1940-41. The suggestion made in these pages is that another superlative should be added to the record book: the world's greatest delusion ("brain robbery")—evolution. For one hundred years this theory has darkened the intellect and dazzled the imagination of civilized man, but now at last the true light is dawning.

How is it that such a monstrous error could beguile so many millions of intelligent people for so many decades? The history of forging lends another illustration which may help us to understand. When in 1925 the notorious Alves Reis (later converted, and an outstanding Christian) counterfeited 500 escudo Portugese notes, he arranged that the printing should be done by Waterlows of London. His forgery went undetected for years because *the same plates were used to print the counterfeit notes as had been used to print genuine ones.* Similarly, the great brain robbery has gone undetected (except by very few) for years because the same word "evolution" is used to describe two absolutely distinct ideas—the one genuine, observable, and true; the other spurious and false. Evolution F (for Fact) denotes "adaptation of a species within a species," which nobody would deny; evolution G (for Guesswork) denotes "The origination of species from earlier and totally different forms" (Oxford English Dictionary), which no one has ever observed. Because Evolution F has been proved, innocent students have assumed that Evolution G must be true.

The new *Oxford Biology Readers* of 1973 furnish a good example of this confusion. In No. 55, "Evolution Studied by Observation and Experiment," Professor E. B. Ford discusses the variation and adaptation of primroses, moths, butterflies, and snails. This is evolution *fact*. On the other hand, in No. 1, "Some General Biological Principles Illustrated by the Evolution of Man" by Sir Gavin de Beer, the same word is used to denote man's supposed origination from the lower primates. (Typically, No. 55's cover illustration is photographic; No. 1's is an artist's "reconstruction" of an ape-woman. One is reminded of another famous reconstruction—that of Hesperopithecus ("Western Ape") in the *Illustrated London News,* June 24, 1922. He was later discovered to be an extinct pig!) This is evolution guesswork. But because Professor Ford's kind of evolution and Sir Gavin's have become inextricably confused in most people's minds, the counterfeit word has passed for genuine and civilization has fallen victim to the greatest brain robbery of all time.

No originality is claimed for these pages. Bigger and better books on the same subject have been written before. But since evolution guesswork continues to ravage our schools and wreck the faith of many, it is hoped that this small volume may cause teachers, and trainers of teachers, to think again.

1

PUZZLED PUPILS

"Mommy, if God did not mean what He said, why did He not say what He *did* mean?"

The little girl's question highlights a problem that has faced every teacher of Genesis over the past hundred years. From time to time theories "harmonizing" science and Scripture have been put forward by brilliant scholars and scientists, only to be shot down by equally brilliant scholars and scientists of the next generation, or even the next decade. Meanwhile the Christian public has been left to wander in the dark, vaguely hoping that, because some scientists are Christians, science does not *really* contradict the Bible. By parity of reasoning one might conclude that, because many Hindus are scientists, science does not really contradict Hinduism, although the Vedas teach that the moon is approximately 150,000 miles higher than the sun and shines with its own light, that the earth is flat and triangular, and that earthquakes are caused by elephants shaking themselves under it! No, the Christian conscience will not long rest satisfied with such questionable logic. That is why the wheel has now turned full circle, and an increasingly large number of people are beginning to wonder whether the "scientific" account of origins is the last word.

We live in an age of intense specialization. Knowledge has become so minutely subdivided that even at Britain's Royal Society, so it is said, not more than 10 percent of the audience understands 50 percent of any lecture! "Expertism" reigns, and a man's opinion on a subject is little valued unless he has graduated through the academic disciplines in that branch of learning. Now, unfortunately the Bible, particularly Genesis, touches on a great number of sciences: anthropology, zoology, archaeology, and philology, to mention but a few. And who can claim to be an expert in more than one or two of these? Very few theologians have engaged in scientific research, and very few scientists have done more than dabble in Hebrew. Thus the interpretation of Genesis may be pronounced hopeless and abandoned at the start, because life is not long enough for anyone to grasp fully the intricacies of every Bible-related subject.

There is, however, another approach, indicated in Scripture itself. Almighty God wills that all men should be saved and come to the knowledge of the truth (1 Timothy 2:4); He caused the Old Testament to be written for *our* learning and *our* admonition (Romans 15:4; 1 Corinthians 10:11); Christ said that till heaven and earth pass away, not one smallest part of any letter of any word of the Law (which includes Genesis) shall become obsolete. He also told some Hebrew scholars:

"Ye do err, not knowing the scriptures, nor the power of God" (Matthew 22:29, KJV). All this suggests that the key to Scripture (once translated) is a combination of *study* of related Bible passages, and *faith*. When these two are aligned, the lock opens. Conversely, when study and/or faith are lacking, God's secrets remain hidden, and the mind (however learned) gropes in darkness. The above-

12

quoted verses also suggest that the God of the Bible would reveal the history of creation with such simplicity and clarity that the man in the street of Tokyo, Timbuktu, or Tooting would not be deceived but would really understand from where he came and where he is going.

The only God worthy of mankind's trust and adoration is the God who can accurately describe the world's past, as a basis for predicting the world's future. This is the very challenge which Jehovah makes to the gods of the heathen in Isaiah 41:22-23: "Declare ye the former things, what they are . . . that we may know that ye are gods." The purpose of this book is to show that in Genesis we have an entirely satisfying account of the creation of the world because it fits the facts of empirical science, and no other account can stand in competition with it. Perhaps, after all, God did say what He meant; perhaps he does mean exactly what He has said.

2

FACING THE FACTS

Without this history the world would be in total darkness, not knowing whence it came or whither it goes. In the first page of this sacred book a child may learn more in one hour, than all the philosophers in the world learned without it in thousands of years.

ANDREW FULLER, 1800

Few books in the history of mankind can truly be called epoch-making; few books have radically changed the outlook and thinking of millions. In the past 500 years we might name Copernicus' *Revolutions of the Heavenly Spheres*, Newton's *Principles*, Karl Marx's *Rise of Capitalism*, and Darwin's *Origin of Species* as among the most influential. It may well be that Morris and Whitcomb's *The Genesis Flood* is worthy to be classed with such books. [1]

The main drawback of *The Genesis Flood* is its length, though it had to be long and technical in order to convince the experts who might be expected to look for particular answers to the particular problems of their fields (such as geology and palaeontology). So it is my modest purpose to summarize the arguments by which Whitcomb and Morris

14

arrived at the conviction that the first eleven chapters of the Bible are literally, historically, and scientifically true. Those who are not satisfied with the presentation given here are urged to obtain *The Genesis Flood* and study it in detail.

Let us frankly admit that to an educated Christian the early chapters of Genesis present a harder intellectual problem than any other part of the Bible. Here he finds an account of the origin of the world totally different from what he has been taught at school and in the university, where the immense age of the earth and stars, and evolution, are assumed to be facts as undeniable as H_2O=water or two times two is four. If he consults the *Encyclopaedia Britannica* he will find these words:

"That the records of prehistoric ages in Genesis 1-11 are at complete variance with modern science and archaeological research, is unquestionable."

Second, Genesis 3:17-19 teaches that some drastic change came over the earth as a punishment for man's disobedience. Calvin comments: "The inclemency of the air, frost, thunders, unseasonable rains, drought, hail, and whatever is disorderly in the world, are the fruits of sin." But science teaches that death, disease, famine and drought, thorns and thistles, and "nature red in tooth and claw" have prevailed on this planet for scores of millions of years. The doctrine of the curse has simply dropped out of the thinking of modern philosophers because no fossils have been discovered of straw-eating lions or vegetarian wolves! Once again there seems to be a head-on collision with the Bible.

Third, Genesis 6-9 appears to teach that there was once a year-long Flood covering the whole globe, but in the *Encyclopaedia Britannica* we find this flatly denied:

External evidence (i.e. Geology) recognizes no universal deluge. . . . Genesis preserves not literal history but popular traditions of the past . . . many of the stories (other nations besides the Jews have Flood traditions) may arise from the inundations caused by the far-reaching tidal waves that accompany earthquakes. . . . Whenever flood-traditions appear to describe vast changes on the surface of the globe, these traditions are probably not the record of contemporary witnesses, but the speculation of much later thinkers.

With this agrees the dictum of G. E. Wright in his *Biblical Archaeology:* "The Flood is an exaggeration of a local inundation."

In conclusion, in Genesis 1-11 we are faced with three facts which the Bible appears very clearly to affirm, and which science equally clearly denies:

1. The creation of the universe in six days of twenty-four hours.

2. The curse on the earth.

3. The universal Flood.

It is now my purpose to show how Whitcomb and Morris have proved *(a)* that these doctrines are indeed taught in the Bible, and cannot be evaded; *(b)* that these doctrines are not contradicted by any *fact* of science but only by the theories of scientists who base their whole philosophy on the false premise of uniformitarianism.

DOES THE BIBLE TEACH A LITERAL SIX-DAY CREATION?

Obviously there is no point in defending a doctrine which God has not revealed, and many evangelical scholars would answer no to the above question. Later, we shall examine some of their theories. But first let us look at the positive arguments for believing that the right answer is yes.

1. *The demands of the context.* The first reason for believing that the Bible teaches a literal six-day creation is this: the context demands it. The word for "day" in Hebrew, as in many other languages, is used with a variety of meanings; but in nearly every case it is obvious from the context what is meant. In Genesis 1:5 the word is used to mean, first, day*light,* and second to include the hours of light *and* darkness. It seems very probable, therefore, that a twenty-four-hour day is intended.

It has often been objected that, since the sun was not "made" till the fourth day, the first three days at least cannot have been solar days. To this, Calvin supplies the answer:

> It did not happen by accident that the light preceded the sun and the moon. To nothing are we more prone than to tie down the power of God to those instruments, the power of which He employs. The sun and moon supply us with light: and according to our notions we so include this power to give light *in them,* that if they were taken away from the world it would seem impossible for any light to remain. Therefore the Lord, by the very order of the creation, bears witness that He holds in His hand the light, which He is able to impart to us without the sun and moon.

2. *The use of the word "day."* Second, in nearly every other Old Testament passage where "days" is used with a numeral, it means literal days of twenty-four hours. The only possible exceptions are Daniel 8:14 and 12:11-12; but these chapters are visions, a type of literature entirely different from Genesis 1, which bears all the marks of being "sober history," according to Professor E. J. Young, who was recognized as one of the world's most eminent Hebrew scholars of the '50s and '60s.

There is an interesting parallel to Genesis 1:1—2:4 in

17

Numbers 7, where we read that "the princes offered for the dedication of the altar *in the day* that it was anointed." We might have thought that all offered on the same day, had not the narrative gone on to inform us that they offered on twelve separate days. We have "the first day . . . second day . . . ", etc., exactly as in Genesis 1; and common sense tells us that the word "day" is used in Numbers 7:10 with a comprehensive meaning, while in the rest of the chapter it is used literally to denote a period of twenty-four hours. Similarly, common sense tells us that in Genesis 2:4 "day" is used with a comprehensive meaning, summing up the six individual and literal days of the previous chapter.

3. *The fourth commandment.* Third, God's commentary in Exodus 20: 9-11 (KJV) states that God's working week and man's working week are exactly parallel: "Six days shalt thou labour. . . . For in six days the LORD made heaven and earth, the sea, and all that in them is."

E. J. Young comments: "The fourth commandment actually refutes the non-chronological interpretation of Genesis One." Let us remember, too, that there is no possibility of the Ten Commandments being man's interpretation of God's word. If any part of the Bible is verbally inspired, this must be, since we are told it was the writing of God written with the finger of God on tablets which were the work of God (Exodus 31:18; 32:16). Dr. Marcus Dods wrote in 1900 "If the word 'day' in this chapter [Genesis 1] does not mean a period of 24 hours, the interpretation of Scripture is hopeless."

For at least two generations this commandment has caused acute embarrassment to the friends of Christianity, and glee to her foes. The situation may not inaptly be

18

compared to a hypothetical scene at Greyfriars School where a student monitor is showing a new boy around. Coming to the headmaster's study door, they find engraved upon it in letters of gold the Ten School Rules. The first five rules are:

1. Mr. Topman is the only authorized headmaster.
2. No boy shall draw any picture of the headmaster.
3. No boy shall call the headmaster by any nickname.
4. No boy shall read by moonlight, because the man in the moon might be offended.
5. Every boy shall write to his parents once a week.

New Boy: They all seem reasonable enough, except for number four. What on earth does he mean by that? You don't mean to say the headmaster really believes in the man in the moon? That's just a fairy tale!

Student: Well—er—um—yes, perhaps. But some boys think there must be some deep hidden meaning to the rule. Anyway, there is a long-standing tradition at Greyfriars that we don't read by moonlight.

New Boy: Well, I think it's ridiculous! A head who puts a statement like *that* in the middle of his rules doesn't deserve to have *any* of them obeyed!

Similarly, thousands of Bible readers have dismissed all Ten Commandments as an antiquated tribal code unfit for Twentieth-century man, because of this "totally unscientific concept" annexed to the fourth.[2] How very *unwise* of God to set all morality at risk by thrusting such a bald, bold, ridiculous statement in the middle of His otherwise reasonable laws—ridiculous, that is, *unless after all it is true!*

4. *The interpretation of older commentators.* The fourth reason for believing that the days of Genesis 1 are literal

twenty-four-hour days is this: the vast majority of Jewish and Christian scholars down the ages have believed them to be so. Origen and Augustine, it is true, thought they might represent ages; but since there is nothing whatever in the text of Scripture to support this idea, it died a natural death. I have already quoted Calvin. The Shorter Catechism of 1647 reads:

Q. What is the work of creation?

A. The work of creation is, God's making all things of nothing, by the word of His power, in the space of six days, and all very good.

Thomas Scott's commentary of 1780 usually mentions varying interpretations where they exist, but he says nothing about any possibility of the "days" being other than twenty-four-hour days. Only since the middle of the nineteenth century, when geologists began dogmatically to assert the immense antiquity of the earth, have Christians begun to doubt. Thus Keil and Delitzsh know of other views, but emphatically reject them: "If the days of creation are regulated by the interchange of light and darkness, they must be regarded not as periods of time of incalculable duration, or of thousands of years, but as simple earthly days." Professor S. R. Driver examines and refutes all the attempts to reconcile Genesis 1 with the dogmas of science, and concludes: "Verses 14-18 cannot be legitimately interpreted except as implying that, in the conception of the writer luminaries had not previously existed; and that they were 'made' and 'set' in their places in the heavens *after* the separation of sea and land." Finally, the German scholar Gerhard von Rad in his monumental commentary on Genesis writes as follows: "Unquestionably the days are to be regarded as literal days of twenty-four hours." (Both Driver and von Rad would explain the six-day creation as a mistaken and primitive idea. I call on them only as acknowledged linguistic experts to tell us what the original writer *meant*. Whether or not the original writer's idea was *true* is another question, which is discussed later.)

5. *The failure of modern commentators*. The fifth reason for believing that the days of Genesis 1 are literal days is this: all attempts to explain the early chapters of Genesis as anything other than "sober history" have, sooner or later, been proved inconsistent, incoherent, or erroneous. The "explanation" often poses more problems than it solves, as the following examples will show:

IS IT A PARABLE?

a. Dr. Alan Richardson interprets the "light" of verse 3 as "the light of God's own presence," analogous to Revelation 21:23.[3] But in that case, where was God's glory when "darkness was upon the face of the deep" (Genesis 1:2)? How could the Spirit of God move *without* His own glory? Surely the light must mean *created* light, which can nevertheless exist (as Calvin pointed out) apart from the sun. The glory of Revelation 21 and 22 will be far superior to the present created light, inasmuch as the new heavens and new earth will be far superior to the present heavens and earth.

IS IT SYMMETRICAL?

b. Many writers have tried to discover a "symmetry" in Genesis 1 which would enable them to call it poetry and to interpret it nonliterally and nonchronologically. E. J. Young has shown this view to be untenable[4]; but *The New Bible Dictionary* actually printed a diagram demonstrating that fishes were made on the *sixth* day (the

Bible says the fifth) in order to prove a correspondence
with the creation of the sea on the third day! The comment
here of Keil and Delitzsch is still very much to the point:

"The creation of fish and fowl on the same day is an
evident proof that a parallelism between the first three
days of creation and the last three is not intended and does
not exist . . . the account before us is obviously a historical
narrative."

Is It Poetry?

c. Others have called Genesis 1 a "Hymn of Creation,"
in which we may expect to find poetical metaphors and
spiritual ideas clothed in figurative language. But Hebrew
poetry has a character as definite as Homeric hexameters
or Shakesperian sonnets; its main features are parallelism
and repetition, which we find even in the first "poem" of
the Bible (Genesis 4:23-24). And parallelism and repetition
are conspicuously absent from Genesis 1, so we cannot
call it poetry. Compare, for example, Psalm 33:6 (KJV):

> By the word of the LORD were the heavens made; and all
> the host of them by the breath of his mouth.
> He gathereth together the waters of the sea together as an
> heap; he layeth up the depth in storehouses."

Here each idea is repeated in different words: this is
poetry. Now Genesis 1:3, 9 (KJV):

> And God said, Let there be light; and there was light.
> And God said, Let the waters under the heaven be
> gathered together unto one place, and let the dry land
> appear: and it was so.

No parallelism, no repetition: this is history. (Admittedly
v. 27 is repetitious, but notice that the *same* verb is used

three times. This kind of repetition is hardly ever found in Hebrew poetry.)

IS A HARMONY POSSIBLE BETWEEN GENESIS AND SCIENCE?

d. Many have tried to prove that the geological time scale agrees with the order of creation in Genesis 1, and for years this was confidently asserted by Christian apologists. However, this line of defense also has had to be abandoned (see chap. 3). A recent booklet which attempts to propagate this mistaken idea is *The Origin of Man* by Victor Pearce.[5] He suggests that we have evidence of man's age of innocence in the "First Danubians" of prehistoric central Europe, because they had no weapons or fortifications. Canon Pearce seems to overlook the fact that Adam and Eve were *fallen* creatures when they were cast out of Eden, and before they had any children. They could scarcely have begotten a tribe of "innocent" nomads. In the very next chapter of Genesis we meet murder and the words "slay," "smite," "wound," and "bruise." There was violence enough in the world, it suggests, long before any migrations to Europe, and the Bible clearly teaches that *only* Adam and Eve were ever innocent, and *only* in Eden. All the artifacts of "early man" must have belonged to Eve's posterity, who were fallen, sinful creatures, condemned (as we are) to toil for a living, and inclined (as we are) to fight one another. But probably then, as now, some tribes were less bellicose than others.

IS THE SEVENTH DAY STILL INCOMPLETE?

e. Another theory is that, because no evening or morning are mentioned in connection with the seventh day, therefore God's Sabbath is still continuing after many thousands of years; hence, each of the six days may represent an equally long and indefinite period. But this theory does not fit the words of Scripture, which are: "He rest*ed* on the seventh day. . . . And God blessed the seventh day, . . . because that in it he rest*ed*" (Genesis 2:2-3, ASV). The basis of all theology is grammar, and grammar distinctly tells us here that God's resting was in the past definite tense, not in the present continuous. Exodus 31:17 says the same: "He rested, and *was* refreshed"; so does Hebrews 4:10 (ASV): "He . . . himself also rested from his works, as God *did* [rest] from his." We conclude, then, that God's rest was for twenty-four hours, exactly as man's should be. We must correct our ideas of God from Scripture, not correct Scripture in order to make it suit our ideas of God.

Is There a Gap Between Genesis 1:1 and 1:2?

f. This "gap theory" was popularized by the Scofield Bible and is still believed by many. Verse 2 has been paraphrased, "And the earth *became* waste and void," and this is supposed to refer to some judgment of God on the "pre-Adamite" earth. Because in Exodus 20:11 the word is "made," not "created," Genesis 1 is interpreted as a restoration of the earth, not the original creation.

On linguistic grounds the theory has been refuted both by E. J. Young and F. F. Bruce,[6] but its error will be apparent even to those who have only the King James Version. Can we distinguish between "created" and "made" in Genesis 1? The answer seems to be no, except in verse 1 where "created" must mean "created out of no preexistent materials." Let us examine the other verses where one or other of the words is used:

Verses 6-7: "Let there be a firmament . . . and God *made* the firmament."

Verses 14-16: "Let there be lights . . . and God *made* the two great lights."

Verses 20-21: "Let the waters swarm . . . and God *created* the great sea-monsters."

Verses 24-25: "Let the earth bring forth . . . and God *made* the beast of the earth."

Verses 26-27: "God said, Let us *make* man . . . and God *created* man, in the image of God *created* he him; male and female *created* he them."

Verse 31: "God saw every thing that he had *made*."

Here we may note (1) It is rather hard to maintain that the process of "creating" sea monsters was different from that of "making" the beasts: in both cases there was preexistent material. (2) The verses about man show that the two words must be considered synonymous in the context. (3) If "made" in Genesis 1:31 can include three acts of creation (vv. 1, 21, 27)—as it obviously does—there seems little reason to doubt that in Exodus 20:11 it can do the same.

Genesis 2:7 uses yet a third word, "formed," to describe the creation of man, but a comparison with Isaiah 45:7 shows that all three words are sometimes synonymous. In Genesis 1 and 2 they denote not three different processes but the *same* process viewed from three different angles, for variety and emphasis.

g. The latest and most ingenious attempt to escape from the literal interpretation of Genesis 1 is found in Derek Kidner's commentary. Kidner suggests that the "latent truth" of the chapter (six days equals 15 billion years) was hidden from Moses, just as Daniel did not know the full import of the words spoken to him: "I heard, but I understood not" (Daniel 12:8, ASV), and Peter says that the prophets did not know the time when Christ would suffer (1 Peter 1:10). Will this analogy hold water?

First let us notice that, though Daniel and Isaiah may not have entirely understood what they were commanded to write, they certainly knew they were writing prophecy (Peter says it was *revealed* to them). But in Genesis 1 there is nothing to suggest that we are reading a prophecy. There is nothing to suggest that Moses did not perfectly understand what he was writing about creation, just as he perfectly understood what he wrote about Joseph. The author evidently regarded the "origin" of the heaven and earth in exactly the same way as he regarded the "origin" of Abraham, Noah, and the rest—as literal history of the past.

Second, one important purpose of biblical prophecy seems to be that after the event God's people may be able to show others how He has manifestly fulfilled His own words. This was certainly achieved in the case of Daniel's prophecies—so marvelously that skeptics and higher critics still maintain they were written *after* the event! Isaiah 53, too, was exactly fulfilled in the life, death, and burial of Christ. The apostles could point with utmost confidence to that chapter as proving the extraordinary fact that the Messiah *must* suffer (Acts 26:23); and Apollos "powerfully confuted the Jews, and that publicly, showing *by the scriptures* that Jesus was the Christ" (Acts 18:28, ASV).

Now in the case of Genesis 1, could a modern Christian "powerfully confute" unbelievers by proving from the Scriptures that the scientific understanding of evolution was stated therein thousands of years before Darwin? We might imagine the posters for a public lecture:

THE BIBLE IS TRUE
SIX DAYS = 15 BILLION YEARS!
GEOLOGY CONFIRMS GENESIS,
THOUGH TOTALLY DIFFERENT!

This is indeed *reductio ad absurdum*. As everyone knows, Genesis has been the laughingstock of unbelievers for a hundred years. So far from any "prophecy" in it being fulfilled, almost every word of the chapter has been held up to ridicule as contradicted by the "assured results of geology." So far from anyone coming to believe Genesis through seeing its "fulfillment," millions of high school and university students have been turned aside from reading the Bible because they found in Genesis 1 an impassable *stumbling block* to faith. How then can this chapter be regarded as prophecy, the purpose of which is to *strengthen* faith by an exact correspondence between the prediction and the event?

CONCLUSION

No, if we are to justify the inclusion of Genesis 1 in the Bible, it cannot be as prophecy any more than as poetry, parable, or picture. When God has told us that He made the universe in six days, and in a certain definite order, adding a precise chronology, it seems rather presumptuous to state that "the Bible only answers the question *By whom?* and not the questions *How?* or *When?* was the world made?" We must unflinchingly face the fact (in the words already quoted) "that the records of prehistoric ages in Genesis 1-11 are at complete variance with modern science and archaeological research." If science is right, then the Bible is wrong; if the Bible is right, then science is wrong.

3

THE ORIGIN OF THE UNIVERSE

Where were you when I laid the foundation of the earth?
JOB 38:4 (RSV)

A question that some may be asking is this: On a scientific matter, of what value is the opinion of anyone who is not a scientist? Should we not leave all such investigations to the experts, and meekly accept their conclusions?

To answer this we must ask another question: What is science? By definition it means *knowledge,* which can refer only to facts discovered by observation, experiment, or trustworthy testimony. So long as a scientist tells us what he or others have observed with their own eyes, or what he or others can reproduce by experiment, we shall be satisfied that he is telling us *facts*. We accept that water is H_2O because we are assured that experiments have proved it to be so. But when a man believes something not on the basis of observation or experiment or trustworthy testimony, he ceases to act as a scientist and begins to theorize as a philosopher. For instance, most scientists believe in evolution, in spite of the enormous gaps in the fossil record. They have not observed evolution, neither can they reproduce it by experiments; they believe it has taken place because it *seems* to them more probable than special creation. But this is not true science; it is faith in the unseen.

One of the chief reasons for assigning to the universe an age of about 15,000 million years is the time required for light to travel from the more distant stars. The argument runs:

1. The distance from the earth to star A equals 1,000 million light-years.

2. We can see this star's light.

3. Therefore the light from star A must have begun traveling toward us at least 1,000 million years ago.

4. Therefore star A must have appeared in the sky at least 1,000 million years ago.

But the whole argument breaks down if at step three above we introduce the God of the Bible. "Why could not God in the twinkling of an eye have formed the stars so that their light could be seen from the earth?" asks E. J. Young. God is not subject to the laws He has made for the normal running of the universe.

Suppose some chemists were to fly back in a time machine to Cana in Galilee in A.D. 26, to a house where a wedding is in progress. Analyzing the wine being drunk at the end of the feast, they would report: "This wine appears to be thirty years old, judging by the speed of fermentation which we observe today in ordinary grape juice." Their observation would be perfectly correct. But suppose they go on to say, "This wine *must* be thirty years old, because the speed of fermentation can never be changed." They would be wrong. Now they speak no longer as scientists but as philosophers, denying that God could or would ever break His own speed limit. (See John's Gospel, chap. 2.)

The point will bear repetition. We shall now send our time machine chemist to a hill by the Lake of Tiberias in A.D. 27. Here he is asked to analyze some barley bread. The report might be: "This bread appears to have been made from barley flour mixed with water and yeast; the flour was produced by grinding barley corn which had grown for six months and had then been cut, threshed, and stored," But if he further asserts that the bread could not have been produced in any other way, the crowd of 5,000 men would retort: "We saw Jesus of Nazareth make the bread with His own hands in (practically) no time. Almighty God is not dependent on the methods which He usually employs. Keep your chemistry for the ordinary processes of life, but don't attempt to analyze miracles!" (See John 6.)

Similarly, as long as the astronomer is content to record present processes, distances, and movements, we may respect and admire his marvelous instruments and complex calculations. But if he proceeds to tell us that God never could or would have acted otherwise than in accordance with processes and speeds *now* observed, we may justly point out that this is not an objective statement of fact but a mere opinion, no less fallible than any other human opinion.

In Hebrews 11:3 we read: "Through faith we understand that the worlds were framed by the word of God, so that things which are seen were not made of things which do appear." The miracle spoken of here was even greater than either of the two mentioned above. Christ incarnate, the Redeemer, used water to make wine and bread to make more bread, but Christ transcendent, the Creator, used *no* preexistent material to form the universe as we now see it. Genesis 1 shows us a fully running world, with an apparent age, created in six days. Just as Christ at Cana compressed into one second the process of thirty years, so in one day He flung the stars billions of light-years into space, at the same time causing their light to fall upon the earth. How do we know? Because He says so!

Still pursuing this question of "apparent age," let us send our time machine man (a physiologist this time) to the Garden of Eden on the sixth day of creation. Applying his height and weight charts to Adam and Eve, he might deduce that they appear to be fully developed adults about twenty-five years old. If he goes on to say that they *must* have been alive for twenty-five years, that they must have been born as he was born, we have a right once more to reply: "Now you speak not as a scientist but as an unbeliever. God says He made Adam and Eve as full-grown adults; and God can do *anything*." So it is in regard to the stars; they all have an "apparent age" which the scientist can measure. But to assert that this is their *actual* age—that they must have been "born" in this way or that way millions of years ago—is to dogmatize quite outside the realm of science.

What Is Faith?

When Noah was told that a Flood was coming he did not start measuring the quantity of water in the atmosphere or the rainfall of Mesopotamia. He did not work out scientifically how such a thing could happen; he simply believed the word of God. When Joshua was told that the walls of Jericho would fall down he did not consult a seismograph or study the statics and dynamics of Canaanite engineering; he simply believed the word of God. So, surely, according to our text (Hebrews 11:3) faith in creation means a simple acceptance of the statement: "God made two

31

great lights . . . he made the stars also. . . . And the evening and the morning were the fourth day" (Genesis 1:16-19, KJV).

Looking more closely at "faith" in Hebrews 11, we shall find that in every case it was exercised in regard to matters of which the believer had no previous experience. Enoch had never seen anyone "disappear" to be with God; Noah had probably never seen rain (Genesis 2:5); Moses had never seen the Red Sea divide. They believed in the "impossible" because God said it would happen. Similarly, creation "out of nothing" has always seemed impossible to unregenerate man, from the most primitive to the most highly civilized. The Thlinkit Indians of North America believed that a creator-hero stole the sun, moon, and stars out of a box, and hung them up to illuminate the earth. Plato taught that matter is eternal, and that all visible things came out of other visible things. This shows the vital importance of Hebrews 11:3; only by personal faith in a personal Creator will anyone ever believe what the world has always deemed impossible—that "it pleased God to create, or make out of nothing, the world and all things therein, whether visible or invisible, in the space of six days" (Shorter Catechism).

4

"A THOUSAND YEARS AS ONE DAY"

We need to remember that limitless time is a poor substitute for that Omnipotence which can dispense with time. The reason the account of creation given here is so simple and so impressive is that it speaks in terms of the creative acts of an omnipotent God, and not in terms of limitless space and infinite *time and* endless *process.*

PROFESSOR O. T. ALLIS, *God Spake by Moses*

In the early days of the Bible-science controversy, Christian apologists often had recourse to 2 Peter 3:8 as an explanation of the six days of creation. They maintained that the "record of the rocks" tends to confirm the biblical order. Genesis 1, they thought, could still be interpreted as a chronological *sequence* even though each "day" must represent many million years. There still remained the perplexing problem of the heavenly bodies, which are said to have been "made" on the fourth day; but this was solved by interpreting "made" as "made to appear," notwithstanding the very slender linguistic evidence upon which this translation is based ("That the heavenly bodies were made on the fourth day, and that the earth had received light from a source other than the sun, is not a naive conception but a plain and sober statement of the truth,")[1]

However, a glance at the geological time scale as given in the *Encyclopaedia Britannica*[2], alongside the Genesis sequence, reveals the impossibility of thus reconciling science with Scripture:

System & Period	Distinctive Records of Life	Began (Millions of Years Ago)	Genesis
	1. CENOZOIC ERA		
Quaternary	Early Man	2+	Day 6
Tertiary	Large carnivores	10	Day 6
	Whales, apes	27	Day 5, 6
	Large browsing animals	38	Day 6
	Flowering plants	55	Day 3
	First placental mammals	65-70	Day 6
	II. MESOZOIC ERA (total 155m. years)		
Cretaceous	Extinction of dinosaurs	130	
Jurassic	Dinosaurs' zenith		
	primitive birds		Day 5
	first small mammals	180	Day 6
Triassic	Appearance of dinosaurs	225	Day 6
	III. PALAEOZOIC ERA (total 345m. years)		
Permian	Conifers abundant		Day 3
	Reptiles developed	260	Day 6
Carboniferous			
upper	First reptiles		Day 6
	Great coal forests	300	Day 3
lower	Sharks abundant	340	Day 5
Devonian	Amphibians appeared		
	fishes abundant	405	Day 5
Silurian	Earliest land plants		
	and animals	435	Day 3, 6
Ordovician	First primitive fishes	480	Day 5
Cambrian	Marine invertebrates		
	(shellfish etc.)	550-570	Day 5

IV. PRE-CAMBRIAN ERA
(total 2920m. years)

No known basis for systematic division	Plants and animals with soft tissues	3490	Day 5
	* * * * * *		
	Origin of the Earth	4500(?)	Day 1
	Origin of Solar System	?	Day 4

* * *

In the above table note that (1) the grand divisions are *not six but four*. Even if we include the era before the earth was "born," that still makes only five divisions; (2) *no two divisions are the same length,* and the first (pre-Cambrian) is forty times as long as the last (Cenozoic). It is hard to see how these four or five periods vastly differing in length could be adequately represented by six equal "days"; (3) the order of events coincides at one or two points, but in general is *very different* in the two columns.

Thus it seems that Kidner's optimism—"a remarkable degree of correspondence can be found between this sequence and the one implied by modern science"[3]—is scarcely justified by the facts. The true solution may lie in another direction: "This uniformitarian, evolutionary scheme of historical geology is basically fallacious"[4]; and "the geological column is an artificial composite affair, with the strata arbitrarily arranged according to the nature of their fossil content to tell the story of evolution . . . there is no spot on earth to which one can go and see more than a few thousand feet of stratified rocks. And in not one of these places can the evolutionary story of any animal or plant be seen."[5]

What, then, does the apostle mean by "one day is with the Lord as a thousand years, and a thousand years as one day" (2 Peter 3:8)? He is preparing the Christian Church for a long delay before the Saviour returns, and with prophetic insight he suggests that it may extend even to (a few) thousands of years. He refers to the *literal* millennia of *human* history, not to the fabulous aeons of prehistory. Bible words do not, like mathematical symbols, possess a fixed value, so that one phrase can be picked up from any passage and dropped into any other, while still retaining its identical meaning. To toss 2 Peter 3:8 into the middle of Genesis 1 is about as sensible as to affirm that Matthew 27:63 means, "After three thousand years I will rise again"!

In passing we may note that the error of scientists in regard to the *past* age of the universe leads them into an equally serious delusion concerning the *future*. Fred Hoyle, ex-Professor of Astronomy at Cambridge, states:

> As the sun steadily grills the earth it will swell . . . until it swallows the inner planets one by one: first Mercury, then Venus, then possibly the earth. This particular idea of the New Cosmology seems to fit in well with mediaeval ideas of Hell . . . after 5000 million years the oceans will boil; after 50,000 million years our galaxy will stop growing. Then what? I should very much like to know.

He concludes:

> The cosmology of the Hebrews is only the merest daub compared with the sweeping grandeur of the picture revealed by modern Science. This leads me to ask the question: Is it reasonable to suppose that it was given to the Hebrews to understand mysteries far deeper than anything we can comprehend, when it is quite clear that they were ignorant of many matters that seem commonplace to us? No, it seems to me that religion is but a desperate attempt to find an escape from the dreadful situation in which we

36

find ourselves . . . we still have not the smallest clue to our own fate.[6]

Professor Hoyle can afford to be dogmatic in his "scientific" predictions of the future because he knows that the whole civilized world has accepted the "scientific" explanation of the past. But it may be hoped that Christians who read his words will realize that, once we abandon the literal interpretation of Bible history, *there is no limit to unbelief*. If "day" does not mean day in Genesis 1, there is no reason why "hell" should mean hell or "heaven" mean heaven in Revelation 21. The children of this world are wiser in their generation than the children of light; they clearly perceive that if Genesis be not a true account of the beginning of the world, there is no reason why Revelation should be considered a true account of the end of the world. The whole Bible stands or falls together, and in the eyes of the world it has fallen, the basis smashed to smithereens. But how firm a foundation has been laid by the new cosmology for the new theology and new morality!

5

SCIENTIFIC PREJUDICE

If you have hitherto disbelieved in miracles, it is worth pausing a moment to consider whether this is not chiefly because you thought you had discovered what the story is really about?–that atoms and time and space . . . were the main plot? And is it certain you were right?

C. S. Lewis, *Miracles*

It is now time to turn our attention to the men who have produced theories of the origin of the universe, and inquire what leads them to deny the validity of the Genesis record. Such scientists are known as "cosmogonists." At this point an illustration may help. Suppose an ardently patriotic and anti-American Frenchman were to write a history of aviation. Having traced the story back to Bleriot's cross-English Channel flight in 1909, he wishes to maintain that the French were the first to invent a flying machine. Therefore he sits down at his drawing board and sketches a number of imaginary planes which *might* have preceded Bleriot's. Being an expert mathematician and engineer, he can argue in a very plausible way that each of these hypothetical machines is an advance on the previous one, that each of them obeys the laws of aeronautics, and that therefore the evolution and invention of the first flying machine by the French is certainly possible. If it is possible, then it is probable. If probable, then practically certain. By repeating his claims over and over again, with an impressive display of mathematical detail, our Frenchman might succeed in convincing his fellow countrymen (many of whom might share his anti-American prejudice) that the honor and glory of inventing the first airplane are due to the French! In fact, of course, the first flight was made by the American Wright brothers at Kitty Hawk, North Carolina, on December 17, 1903. The historical record is unimpeachable, and no amount of theorizing, speculation, or drawing-board pictures, can ever disprove it.

Now I believe that the modern cosmogonist is not unlike that Frenchman. Thoroughly indoctrinated with the theory of evolution, he rules out all possibility of Genesis 1 being literally true.

He settles down at his astronomical "drawing board" and produces, with impressive mathematical detail, a series of hypothetical explosions and contractions, gyrations and convolutions, which *might* have led to the formation of the heavenly bodies as we now observe them. He writes a book on the subject, which is soon followed by another, then another, as he increasingly persuades himself and others that his drawing-board "creations" *actually happened!* After ten or twenty years fellow citizens of his world (many of whom are equally unwilling to accept God's revelation) forget all about the historical record in Genesis. They come to see the drawing-board theories first as possible, then probable, then practically certain. If other equally learned astronomers produce other and totally different designs, that matters little. Somehow or other, science has "disproved" the Bible!

39

Here we may note a strange anomaly. Science itself—and not least astronomy—depends on historical records. Edmund Halley, who assisted Isaac Newton, discovered the comet subsequently called "Halley's Comet" by comparing the written records of twenty-four comets observed from 1337 to his own day (1704). He found that four comets had moved in practically identical paths (1456, 1531, 1607, and 1682, i.e., at intervals of about 75½ years). He therefore correctly assumed that the four appearances belonged to one body, whose return might be expected about 1758. This prediction was justified by the event; the comet returned in 1759 and again in 1835 and 1910. The important thing to notice is that every one of these appearances, apart from the last (which was photographed), is known to us now only *by written testimony of reliable witnesses*. This is real science: knowledge based upon observation and the observer's testimony.

6

WHITCOMB'S "ORIGIN OF THE SOLAR SYSTEM"

Let us now take a brief glance at the astronomical "drawing board" and see whether the designs on it actually fit the facts. Let us remember that even if they did perfectly explain how the universe *might* have evolved, that would be still very far from proving that it did indeed so evolve. The Frenchman's theoretical airplanes, even if each one could be built and made to fly, would never disprove that the Wright brothers came first. But the interesting thing is that not one of the theories of "cosmic evolution" does fit the facts. Professor John Whitcomb in his *Origin of the Solar System* has pointed out that there are at least *nine insuperable difficulties* facing the evolutionary cosmogonist. Of these I here list seven, in the form of questions:

1. If the planets were thrown off from the sun by centrifugal force, why is the sun rotating slower than any of them? An experiment with a top will soon show that small particles thrown off it will lose speed much quicker than the top itself.

2. How is it that Uranus and Venus rotate (on their own axes) in the direction *opposite* to that of the other seven

41

planets? If all nine "evolved" from the sun, it would seem inconceivable that this should be so.

3. How is it that eleven out of the thirty-two satellites of the planets revolve in a direction opposite to that of the revolution of the planets around the sun? The eleven are: four out of Jupiter's twelve; Phoebe, the outermost of Saturn's nine moons; the five moons of Uranus; and Triton, the inner of Neptune's two satellites.

4. Why is it that, whereas the sun rotates much *slower* than its planets, each satellite-owning planet (except the earth) rotates *faster* than its satellites?

5. Many school children are still being taught that the moon was pulled out of the Pacific Ocean 4,000 million years ago, although this theory was exploded by Harold Jeffreys in 1931. Recent research has shown that the moon's density is only two-thirds of the earth's. If the two bodies were originally one, why are they now so differently composed?

6. How did the earth come to have such a huge proportion of heavy elements (iron, nickel, magnesium) compared with the sun, which is 99 percent hydrogen and helium? Professor Hoyle says, "Material torn from the sun would not be at all suitable for the formation of the planets as we know them"[1] (Here I call on Professor Hoyle as a witness to demonstrable *facts* in the immediate *present*. This is very different from accepting his *theories* about the earth's unknowable—apart from Genesis—past and unpredictable—apart from Revelation—future.)

7. The American cosmogonist George Gamow believes that all the elements evolved out of each other by "neutron capture," beginning with hydrogen (atomic weight 1) up to the heaviest elements. Unfortunately for this theory there are two gaps in the chain: there is "no stable atom of mass

5 or mass 8. The question then is: How can the build-up of elements by neutron capture get by these gaps? The process could not go beyond helium 4, and even if it spanned this gap it would be stopped again at mass 8 . . . this basic objection to Gamow's theory is a great disappointment."[2]

In concluding this section I feel bound to draw attention to a fact hinted at before: that some scientists have a prejudice against ascribing to God any credit at all for creation. Not only do they deny that He made everything in six days; they deny that He made anything! Harlow Shapley, Professor Emeritus of Astronomy at Harvard University, makes the following remarks in *Beyond the Observatory:*

"Formerly the origin of life was held to be a matter for the Deity to take care of; it was a field for miracles and the supernatural. But no longer.

"We now believe that all the scores of kinds of atoms have evolved naturally from hydrogen, the simplest of atoms." ("Believe" is the right word to use here; it is purely a matter of faith in the unseen, because the evidence just does not exist.)

"All these (microbes, whales, sequoias, bacteria, bees, sponges) are products of the cosmic processes that evolve atoms, biological cells, plants, animals, and mankind."

The climax of Professor Shapley's new theology is reached in these words: "To me it is a *religious* attitude to recognise the wonder of the whole natural world, not only of life . . . why not *revere* also the amino acids and the simple proteins from which life emerges?"[3] Compare this with Romans 1:21-25 (NEB): "Knowing God, they have refused to honour him as God, or to render him thanks . . . they have bartered away the true God for a false one, and have offered *reverence and worship to created things* instead of to the Creator, who is blessed for ever."

Thus every theory of cosmic evolution so far propounded is confronted with insuperable difficulties. Even if all these were to be solved, no one could ever prove that things actually happened as they are supposed to have happened. There is no reason for doubting, but every reason for believing that God created the stars and solar system exactly as they are today—partly regular and partly irregular, so that man might be forced to confess: "Marvellous are thy works . . . such knowledge is too wonderful for me" (Psalm 139:14, 6). This was certainly the belief of Isaac Newton, perhaps the greatest scientist of all time. And the disastrous effects of unbelief are shown by the virtual idolatry of a great astronomer.

Historical records are "scientific" in the deepest and truest sense. The historical record of God's creating the universe in six days was written by the finger of God on tables of stone, the accuracy of the whole book being vouched for by Christ Himself.

7

GALILEO, DARWIN, AND "INTERPRETATION"

Before proceeding further, it may be well to deal with an objection which is often raised: "Galileo was right and the Church was wrong. Therefore anyone who interprets the Bible so as to make it contradict the findings of science is bound to be wrong, and scientific opinion is bound to be right. It's all a matter of interpretation." Let us carefully consider whether the two cases are really parallel.

1. The only words of Scripture which *seemed* to contradict Galileo were three verses (in Psalm 93:1; 96:10; 104:5), which are *poetry*. Nowhere in the historical books is there one word that denies the earth's rotation.

Darwinism, on the other hand, contradicts nearly everything written in the first eleven *chapters* of Bible *history:* it denies instant creation, the fixity of species, the special creation of man and woman, the Fall, the curse, the universal Flood, the miraculous confusion of tongues, and the young age of the earth. All these doctrines are stated in plain prose, and many are confirmed by New Testament references.

2. Later commentators have had no difficulty in showing that the words which got Galileo into trouble ("the

world is stablished that it cannot be moved'') are in a sense very true. We speak of a man "keeping his place" on a football field when we mean that his position remains the same *relative to the other players*. Similarly, the earth can be said "not to move" out of its orbit and position *relative to the other planets* and the sun. This is a perfectly satisfying explanation and has been generally accepted.

On the other hand, no one has yet put forward, even after a hundred years of discussion and controversy, any acceptable harmony of Genesis and Darwinism. Somehow those 299 verses are terribly resistant to "bending"!

3. The Copernican system is *demonstrable* here and now and explains all the phenomena. But evolution leaves a great many phenomena unexplained; it is at best a theory "unproven and unprovable," demanding for its acceptance "faith of the highest kind: faith in the fossils which have never been found, faith in the embryonic evidence which does not exist, faith in the experiments which refuse to come off, faith unjustified by works!"

So here is the difference between Darwin and Galileo: Galileo set a demonstrable *fact* against a few words of Bible poetry which the Church at that time had understood in an obviously naive way; Darwin set an unprovable *theory* against eleven chapters of straightforward Bible history which cannot be reinterpreted in any satisfactory way. The "parallel" is not parallel at all. This leads us on to the subject of interpretation.

In the Bible the word "interpret" is used only in connection with foreign languages, parables, prophecies, dreams, and visions. In the Old Testament it is hardly used outside Genesis 40-41 and the book of Daniel. Two important points should be noted:

1. In every passage where things are reported contrary

to natural science (sun and moon bowing down to Joseph, lean cows eating fat cows, a lion with eagle's wings), it is clearly stated that it was a dream or vision.

2. In every passage the interpretation of the dream or vision is given (at least in general terms, e.g., Daniel 11). In Genesis 1-11, however, there is nothing to suggest that it is a dream or a vision or symbolic; and nowhere in the Bible do we find any interpretation of these chapters. May we not assume that from God's point of view they are *not* contrary to (super) natural science, which is the only true science of the universe? Genesis 1-11 no more requires interpretation than the statement "Balaam saddled his ass" or the football scores in your morning newspaper.

8

IS THE EARTH "VERY GOOD"
OR "CURSED"?

Belief in words is the basis of belief in thought.
BISHOP WESTCOTT

When I use a word, it means just what I choose it to mean.
HUMPTY DUMPTY

*If God's moral judgement differs from ours so that our
"black" may be His "white", we can mean nothing by
calling Him good.*
C. S. LEWIS

THE VIEW OF OLDER COMMENTATORS

For eighteen centuries Christians took Genesis at its
face value, believing that it required no more "interpretation" than any other historical book. They accepted that
the original animal world was created vegetarian (1:29-30);
beasts did *not* prey upon one another, so the balance of
nature must have been kept by a process which we do not
now observe.

They found this view confirmed by the prophecy in
Isaiah 11:6-9, (ASV): "The wolf shall dwell with the lamb,
and the leopard shall lie down with the kid . . . the lion shall

eat straw like the ox. They shall not hurt nor destroy in all my holy mountain.''

Calvin comments:

> Isaiah describes the order which was at the beginning before man's apostasy produced the unhappy and melancholy change under which we groan. Whence comes the cruelty of beasts . . . ? There would certainly have been no discord among the creatures of God, if they had remained in their original condition. When they exercise cruelty towards each other . . . it is an evidence of the disorder which has sprung from the sinfulness of man . . . if the stain of sin had not polluted the world, no animal would have been addicted to prey on blood, but the fruits of the earth would have sufficed for all, according to the method which God had appointed.

Romans 8:20-22 (ASV) seems to refer to the same event: "The creation was subjected to vanity . . . the whole creation groaneth and travaileth in pain together until now."

Genesis 3:16-19 also appears to teach that if Eve had not sinned, childbearing would have been comparatively painless; thorns and thistles would never have existed; man would have lived without toil, on fruit plucked from trees; and the human race would have been immortal. Calvin comments:

> The Lord determined that His anger should, like a deluge, overflow all parts of the earth, that wherever man might look, the atrocity of his sin should meet his eyes . . . whatever unwholesome things may be produced, are not natural fruits of the earth, but are corruptions which originate from sin . . . all the evils of the present life have proceeded from the same fountain. Truly the first man would have passed to a better life, had he remained up-

right; but there would have been no separation of the soul from the body, no corruption, no destruction, and no violent change. It is credible that woman would have brought forth without pain, if she had stood in her original condition.[1]

THE VIEW OF MODERN COMMENTATORS

All this is flatly denied by modern science. The dinosaurs, most ferocious of all carnivores, are supposed to have roamed the earth, tearing their victims, 180 million years before man appeared. No female skeleton has ever been discovered better adapted for childbearing than modern woman's, "prehistoric" man evidently used weapons and agricultural implements, and fossil skeletons presumably belonged to *mortal* men. So there is not one shred of tangible evidence that the age of innocence ever existed. Science also asserts that the ancestry of bacteria, cancer, scorpions, and other noxious creatures can be traced back or assumed far beyond the origin of *Homo sapiens*. Which is right, science or the Bible?

The neoevangelical answer is that science is right, and the older commentators were wrong in their interpretation. The all-pervading climate of uniformitarianism has forced even reputable scholars to declare that the world *as now constituted* is "very good." Included in this "goodness" are not only thorns and thistles, cancer and leprosy, scorpions, earthquakes, childbirth without chloroform and death, but also the rack and the thumbscrew, Sodom and Gomorrah, Belsen and the atomic bomb. "If there were no evil men, many good things would be missing in this universe."[2] If irresistible logic drives Dr. Ramm to such a desperate conclusion, one may reasonably ask whether his premises are sound. For, if words can be used in the Bible with a sense opposite to that which they bear in human communication, then revelation from the mind of God to the mind of man is impossible, and the whole Christian faith is reduced to confusion. In the following chapters it is shown that Dr. Ramm's premises (uniformitarianism, a local Flood) are *not* sound; here we shall take a closer look at the words of Scripture.

What the Bible Says

Whitcomb and Morris show that the neoevangelical view involves wresting Scripture to make it mean what no ordinary reader would ever take it to mean.[3]

1. Romans 5:12, 14 (KJV) states that "by one man sin entered into the world, and *death by sin. Death reigned from Adam*" (not before Adam).

2. Genesis 3:17-18 does not say, "Cursed art *thou* from the Garden; from henceforth thou shalt be removed to the thorns and thistles"; it does say, "Cursed is *the ground* for thy sake . . . thorns also and thistles shall it bring forth to thee," and Genesis 5:29 stresses the same fact—the Lord's curse upon *the ground*. The inescapable inference is that thorns, thistles, and all other "unwholesome things" began to grow only *after* the Fall.

3. Eve's punishment is obviously beyond the scope of scientific investigation, since *only* Eve had a perfect body, and it must have been miraculously changed to cause her and all her daughters to suffer greatly multiplied pain in childbirth. It is common knowledge that most animals appear to suffer far less than humans in parturition, but no evolutionist has yet explained why.

4. C. F. Keil comments on Genesis 3:14:

"If these words are not to be robbed of their entire meaning, they cannot be understood in any other way than

51

as denoting that the form and movements of the serpent were altered, and that its present repulsive shape is the effect of the curse pronounced upon it."[4] And if God could change the shape of the serpent, there is no reason to doubt that He changed the eating habits of other animals at the same time. In the garden they were brought to Adam and were under his dominion (1:28); but after the Fall, the beasts are potentially man's murderers (Genesis 9:5).

CONCLUSION

The age of innocence will never be "discovered" because no *physical* record could possibly exist of a world that lasted only (perhaps) a few days or months. If a marriage breaks up after ten years, there may be nothing to show there was ever a honeymoon—nothing except the family diary. And Genesis is the first chapter in the diary of God's family. To refer to a previous illustration, in the days before photography the only possible record of Halley's Comet was *human testimony*, which was proved true when the same comet reappeared after the stated interval of years. Similarly, the only possible record of the age of innocence is the human testimony (inspired by God) which we find in the Bible, and which will be proved true when the second paradise appears, a kingdom in which "death shall be no more; neither shall there be mourning, nor crying, nor pain, any more. And there shall be no curse any more" (Revelation 21:4; 22:3).

All scientific and theological objections to the doctrine of the curse are based upon (1) the idea that God *could not* have created plants and animals different from what they are now. (But if God could not do this, He is no longer God.) (2) The idea that God *does not* interfere with the course of nature. (But this theory is contradicted throughout the Bible.) (3) Denial of the universal Flood, which, as will be shown in subsequent chapters, can well account for all the fossils of dinosaurs and other flesh-eating animals: they lived and died *after* the curse.

Surely it is more reasonable to accept the older commentators' interpretation, which science can never disprove:

1. That the original creation was literally paradise, a perfect environment for man and woman to live in perfect health, perfect happiness, perfect holiness, and fellowship with God.

2. That all our physical troubles are *not* good in the sight of God, but have been "built in" to "this present evil world" (Galatians 1:4) to remind us of our disobedience and continuing sinfulness. Christ healed the sick, fed the hungry, calmed the storm, and raised the dead, to demonstrate that death, storms, hunger, and sickness are due simply and solely to man's sin; and that He, as Redeemer, has power to deliver not only from its penalty and pollution, but also from its effect and environment.

9

THEISTIC EVOLUTION AND GENESIS 2:7

The doctrine of Evolution, if consistently accepted, makes it impossible to believe the Bible.

T. H. HUXLEY

Many commentators and scientists today teach that "the text of Genesis would by no means disallow"[1] God's shaping man by a process of evolution. They quote as parallels Job 10:8, "Thine hands have made me and fashioned me" and Psalm 119:73 (KJV). The argument runs: obviously Job and David were born in the normal way, yet they speak as if God had formed them directly. Therefore, Adam too might have been born in the normal way (but of a subhuman mother), although Scripture seems to teach his direct creation from the dust of the ground (Heb, *adamah*).

However, I believe that a careful study of the text in its context will show that this interpretation is impossible, for the following reasons:

1. Job's speeches and David's psalms are not history but *poetry*. Job 10:10 says: "Hast thou not poured me out as milk, and curdled me like cheese?" which can scarcely be

intended literally. In Bible exegesis, almost everything depends on determining the literary genre to which a passage belongs, and we believe good reasons have been given for regarding Genesis as "sober history."

2. The Hebrew words for "living soul" in Genesis 2:7 are exactly the same words as those translated "living creature" in 1:20, 21, 24. Therefore, if Adam's body evolved from the lower animals, he would have already been a "living soul/creature" long *before* God breathed into him the breath of life; and the breath of God would (based on this interpretation) have accomplished nothing! But Moses does not say, "God breathed into a living creature the breath of (spiritual) life, and the brute became a spiritual creature." He does say:

"God formed man [Adam] of the dust of the [inanimate] ground [*adamah*], and breathed into his nostrils the breath of life; and man became a *living* soul" (2:7). Before God breathed into him, Adam lacked one thing that the animals possessed: physical life. Therefore Genesis 2:7 cannot be interpreted as God's transformation of a living brute into a living man.

3. In Genesis 3:19 we read: "In the sweat of thy face shalt thou eat bread, till thou return unto the ground; for out of it wast thou taken: for dust thou art, and unto dust shalt thou return."

Let us paraphrase this according to the evolutionary interpretation of Genesis 2:7, and see whether it makes sense:

". . . till thou return unto the sub-human brute; for out of him wast thou taken: for a sub-human brute thou art (physically), and *unto a sub-human brute thou shalt return!*"
This may throw light on the doctrine of transmigration of souls, but more probably it highlights the impossibility of

interpreting "ground" or "dust" as a metaphor for the lower animals.

4. Christ and the apostle Paul refer to these chapters as literal historical fact:

"From the beginning of the creation [not after many millions of years] God made them male and female" (Mark 10:6, KJV).

"Woman was made out of man" (1 Corinthians 11:12, NEB).

"For this cause [i.e., because woman was taken out of man] shall a man leave his father and mother, and shall cleave to his wife" (Matthew 19:5).

It is difficult to see how these verses can be reconciled with the theistic evolutionary picture of God breathing His image into a pair of anthropoids. On the other hand, these verses are exactly what we would expect if Christ and Paul accepted the *prima facie* interpretation of Genesis 1 and 2: that Eve was formed from a rib of Adam, who was formed from the ground on the sixth day of creation.

THE TIME OF MAN'S CREATION

5. Another verse very hard to reconcile with evolution is Romans 1:20 (RSV): "Ever since the creation of the world [kosmos] his [God's] invisible nature, namely, his eternal power and deity, has been clearly perceived in the things that have been made." Bishop Handley Moule comments: "The Greek scarcely allows the interpretation 'from the framework, or constitution, of the world'".[2] "From the creation of the world" (KJV) means "since the world was created." But how could God's power be perceived without a perceiver? Surely this implies that man must have been present *at the beginning* to praise and adore the Creator for His marvelous handiwork. The New English

Bible is even clearer: "Ever since the world began, God's invisible attributes have been visible to the eye of reason."

But where was the eye of reason before man appeared? In the anthropoid ape? In the brontosaurus? In the trilobite?

If Adam was created on the sixth day, only two days after the sun, moon, and stars, Paul's words are easily intelligible. But according to the theory of evolution there were no eyes of any kind to observe God's creation for the first 14,000 million years; animal eyes appeared only (about) 500 million years ago; and the eye of reason (i.e., man's) only *two* million years ago. If this were true one might wonder why Paul did not write: "From the day of *man's* creation, God's power and deity have been clearly perceived, have been visible to the eye of reason"—since the whole thrust of Paul's argument is *man's* failure to glorify God. But Paul did not write this; he did write, "ever since the creation of the *world*" they have been visible. The simple and obvious meaning is that man's eye of reason appeared simultaneously with the rest of the "kosmos"; that man's perception of God's invisible nature has been a fact ever since the world was completed, that is, ever since the sixth day.

Paul's Parallel

6. Finally, let us examine the one New Testament passage (1 Corinthians 15:42-45) which directly quotes Genesis 2:7. Here, if anywhere, we may expect to find the true interpretation. Paul is answering the questions, How are the dead raised? and With what body do they come? First he shows from nature that there are two kinds of body—the seed and the plant—and that the seed must *die* before the plant comes up. Then he continues:

57

"So also is the resurrection of the dead . . . it is sown a natural body; it is raised a spiritual body. If there is a natural body, there is also a spiritual body. So also it is written, The first man Adam became a living soul. The last Adam became a life-giving spirit."

Now, to what event do these last words refer? Paul is not here discussing the incarnation, the virgin birth, or the human life of our Lord. He is discussing one subject only—the resurrection. So it seems very probable that "became a life-giving spirit" refers to that moment of time in which God raised Him from the dead. Then the creation of the first Adam was exactly parallel to the resurrection of the last Adam; just as God breathed into the *lifeless* body of Christ the breath of (spiritual) life, thus making a spiritual body, so He breathed into the *lifeless* body of Adam the breath of (natural) life, thus making a natural body. The analogy is clear and striking.

CONCLUSION

We conclude, then, that New Testament quotations of Genesis 1-3 exclude the possibility of man's origin by theistic evolution. By all the laws of language, Genesis 2:7 can only mean that Adam's body was a special creation, wholly distinct from the animals; that it was literally dead until God quickened it with natural life, just as Christ's body was literally dead until God quickened it with spiritual life. Scientifically there can never be any "explanation" of the resurrection, because the miracle will not be repeated till the return of Christ; and scientifically there can never be any explanation of man's origin, since that too was a once-for-all instantaneous miracle.

10

IS GENESIS THREE A PARABLE?

A miracle's peculiarity is that it is not interlocked with the previous history of Nature. And that is what some people find intolerable. The reason they find it intolerable is that they start by taking Nature to be the whole of reality.

C. S. LEWIS, *Miracles*

Many Christians are still bothered by the talking snake in Genesis 3 and try to explain it away in terms of the book of Revelation. If the word "serpent" in Revelation 12:9 is used symbolically, why should it not be symbolic in Genesis 3 too? The answer lies again in a discrimination of the various types of literature found in the Bible. Revelation is a *visionary* and *symbolic* book, whereas Genesis is a *historical* book. Therefore the language used in Revelation is mostly figurative and metaphorical, while the language of Genesis is mostly straightforward and literal.

Second, we find that throughout Revelation, John expresses spiritual truths in terms of the historical events of the Old Testament. Seven of the plagues of Egypt (water into blood, hail, fire, sores, darkness, locusts, and frogs) are repeated; but we do not therefore conclude that the literal and physical plagues of Egypt never happened. We find "the great city which spiritually is called Sodom," but

59

we do not thereby infer that Lot's Sodom never existed. We find "fire and brimstone" which burn forever, but we do not therefore deny that the cities of the plain were overthrown by literal fire and brimstone. In Revelation 2, "Jezebel" and "manna" are no doubt symbolic, but this does not mean that Ahab's wife was a disembodied spirit, or that the Israelites in the wilderness lived on thin air. Rather, every Old Testament reference in Revelation, though used spiritually in that book, confirms the historical truth of the original narrative.

God's method of teaching (as in the tabernacle) is from the literal, visible, and tangible, to the spiritual, invisible, and intangible. So there seems to be good ground for believing that the tree of life in Eden was a literal botanical tree, which will have its antitype in the spiritual tree of life in heaven; and that the serpent in Eden was a literal zoological specimen which has its antitype in Satan. If a "dumb ass spake with man's voice" to Balaam (as the apostle assures us in 2 Peter 2:16), is it hard to believe that a dumb serpent spoke with a man's voice to Eve? Note also that when Paul refers to this event (2 Corinthians 11:3), he says that the *serpent*, not Satan, beguiled Eve.

Another objection to the literal interpretation of Genesis 3 is based on the prophecy in verse 15: "It shall bruise thy head, and thou shalt bruise his heel." These words, it is said, have a symbolic and spiritual meaning; therefore, the whole passage may be symbolic and allegorical. But, once again, this inference is unwarranted, because it confuses things that differ. Even within one book of Scripture there may be several literary forms—history mingled with prophecy (e.g., Luke 3:1-6), narrative alongside parable (e.g., John 6). But this does not allow us to spiritualize the history or allegorize the narrative. So, while Genesis includes some prophecies and metaphors, especially in the utterances of God, (e.g., "the voice of thy brother's blood crieth unto me", Genesis 4:10), these can in no wise affect the literal historical truth of the narrative portions.

Finally, it has been suggested that Genesis 3 is a parable like Nathan's told to David (2 Samuel 12), in which David's real sin was compared to the rich man's imaginary sin.[1] So the eating of the fruit may be only an imaginary sin representing some greater but unrecorded act of disobedience. Is this interpretation legitimate?

Every parable or allegory in Scripture *either* refers back to some historical event which has already been told in great detail (e.g., David's sin in 2 Samuel 11), *or* it is interpreted (e.g., Isaiah 5:7), *or* it is prophetic (e.g., Matthew 13:33), to be explained by God's working in the history of the Church. So, if Genesis 3 is a parable, it is unique. The event to which it refers is nowhere mentioned; it is not interpreted in any part of the Bible, and the story can hardly be thought of as prophetic.

Then, again, all allegories and parables in both Testaments present spiritual and unfamiliar concepts in terms of natural and familiar truths. But Genesis 3 deals with things absolutely unknown to mortals—a talking snake, nakedness without shame, and life without toil. How then can we be expected to grasp a deeper spiritual truth through a story of the *un*familiar and *un*natural? This was not the method of Nathan the prophet, nor of our Lord.

In my opinion, all such nonliteral interpretations, though put forward with the best of intentions in order to remove intellectual difficulties, serve only to weaken faith in the absolute power of God. Such a faith is our first and greatest requisite for the understanding of Scripture.

11

BIBLICAL ARGUMENTS FOR A UNIVERSAL FLOOD

The greatest derangement of the human mind is to believe because one wishes it to be so.

PASTEUR

We now approach the most important part of our study. Whitcomb and Morris have rightly judged that all interpretation of Genesis 1-11 ultimately depends on the Flood; if it was a local inundation, then the geological time scale, dinosaurs, etc., cannot be fitted into any part of Scripture. On the other hand, if the Flood *was* universal, it adequately explains all the phenomena; and if all the fossils were fossilized after Genesis 6, there can be no scientific or philosophical objections to the idea of a six-day creation. *The Genesis Flood* lists seven biblical arguments for a universal Flood:

1. *The depth of the Flood.* The waters covered the highest mountains to a depth sufficient for the ark to float over them. When they subsided, the ark landed on "the mountains of Ararat," of which the higher peak is 17,000 feet and the lower peak 13,000. Only after the waters had decreased for another ten weeks did the other mountaintops become visible (Genesis 8:5).

2. *The duration of the Flood*. It prevailed for five months, and an additional seven months were required for the waters to subside and the earth to become dry. Such a flood would be quite inconceivable if limited to an area of a few hundred square miles.

3. *The expression "fountains of the great deep were broken up" (for forty days)*. This points unmistakably to vast geological disturbances that are incompatible with the local-flood concept.

4. *The size of the ark*. Its displacement tonnage was about 20,000 tons, and its total capacity 1,400,000 cubic feet. Not until 1884 was any bigger ship built anywhere in the world. Would such a vessel have been necessary for carrying eight passengers and a few Asiatic fauna?

5. *The need for an ark*. If the Flood had been only local and limited, Noah and his family could easily have escaped to another country, just as Lot and his family escaped from the local and limited fire upon Sodom and Gomorrah.

6. *The distribution of the human race in Noah's day*. At least 1,656 years had elapsed from the creation of Adam to the year of the Flood. Considering the longevity and fecundity of the patriarchs, there are good grounds for supposing that the population of the earth in Noah's day numbered millions, that they were widely spread, highly civilized, and possessed good communications. The hundred years while the ark was being built (Genesis 5:32; 6:6) would have allowed plenty of time for news of Noah's preaching to reach the four corners of the world; and in order to destroy *all* flesh (mankind) the Flood would have had to cover the globe.

Evidence of human fossils in Java, China, South Africa,

and western Europe, makes it difficult to assume that men did not migrate beyond the Near East before Noah's Flood.

7. *The apostle Peter's inspired comments.* Three times he refers to the destruction of the ancient *world* (kosmos). This word is used with a limited meaning only twice (Matthew 4:8, possibly; Colossians 1:6, perhaps prophetically); and even if in Paul's view it meant only the Roman Empire, it is obvious that a flood which submerged the Roman Empire of A.D. 60 would have been universal. (We must remember that it is not Noah who says that God destroyed the ancient world, but Peter; and presumably Peter's world was the same as Paul's.)

But the evidence is much stronger yet. In 2 Peter 3:6-7 the apostle contrasts the "world [kosmos] that then was" with *"the heavens and earth,* which are now"; in other words, "kosmos" is the equivalent of our "universe," comprising both heaven and earth. Peter here specifically states that the Flood destroyed (in some sense) *both* heaven *and* earth. This exactly fits *The Genesis Flood* hypothesis that the Flood was caused by precipitation of the water canopy surrounding the earth.

We can scarcely doubt, then, that Peter is telling us that the Flood covered the entire globe and destroyed all mankind except the eight, just as at Christ's return the entire globe will be destroyed by fire, and only those who belong to Him will be saved.

Nongeological Objections Answered

Four of the commonest objections raised to the above interpretation are:

1. The words "all" and "every" in the Bible do not always mean literally all without exception. *The Genesis*

Flood answers this by showing, first, that the context of Genesis 6-9, including the tenor of the entire Flood narrative, demands a literal interpretation of the universal terms ("all" and "every" are used 57 times in these chapters). Second, a flood which rose fifteen cubits above Mount Ararat could not have been anything less than worldwide.

2. How could Noah have gathered and cared for all the animals if two of *every* kind were to be included? *The Genesis Flood* answers this by another question: How do we know that the antediluvian continents, mountains, deserts, climates, and zones were the same as ours? They were probably quite different, and possibly all the continents were joined together by land "bridges." Second, we cannot rule out the supernatural direction of God, who said, "Two of every sort *shall come unto thee*" (6:20). Even now the word "instinct" is used to conceal our ignorance as to how birds migrate to particular places which they have never seen. Third, the size of the ark was ample to accommodate the 35,000 vertebrate animals that would have been required. They would have fitted into 146 two-deck American stockcars, and the ark's capacity was 522 of these. Fourth, it is quite probable that many of the animals "hibernated" during the year-long Flood and needed very little to eat or drink.

3. Where did the rain come from? The answer is: from the water canopy which is described in Genesis 1:6-8 as "the waters which were under the firmament," and which are now mingled with the oceans. (Peter describes the antediluvian earth as "compacted out of water and amidst water" [2 Peter 3:5], that is, as *different* from our present earth.) So the argument that the water now present in the atmosphere would have been insufficient for the Flood, has nothing to do with the case.

4. How could the animals have reached the countries where they now live?

It is by no means unreasonable to assume that all land animals in the world today have descended from those which were in the Ark. In spite of the lack of evidence of marsupials (kangaroos etc.) having lived in Asia, it is quite conceivable that marsupials could have reached Australia by migration waves from Asia, before that continent became separated from the mainland. Comparatively little is known of the migration of animals in the past; but what we do know indicates very clearly the possibility of rapid colonization of distant areas, even though oceans had to be crossed in the process. It would not have required many centuries for animals like the edentates (i.e. armadillos) to migrate from Asia to South America over the Bering land bridge. Population pressures, search for new homes, and especially the impelling force of God's command to the animal kingdom (Genesis 8:17), soon filled every part of the habitable earth with birds, beasts, and creeping things."[1]

Thus every one of the nongeological objections can be answered if we carefully compare the *facts* known to science with the actual words of Scripture.

12

MODERN GEOLOGY AND THE DELUGE

We are indeed a blind race, and the next generation, blind to its own blindness, will be amazed at ours.

L. L. WHITE[1]

Most readers of these lines will have grown up in an educational milieu which assumes the truth of the geological time scale. From childhood we have been awed by pictures of gigantic reptiles with polysyllabic names, and we have imbibed with our mother's milk the "fact" that these monsters roamed the earth innumerable aeons before man. It will therefore be something of a shock to find in *The Genesis Flood* photographs showing that the dinosaurs were *contemporary* with humans, that they walked on the beds of the same rivers at very much the same time. (A more recent publication, *Man's Origin, Man's Destiny,* in 1968 by Professor A. E. Wilder Smith, D.Sc., Ph.D., F.R.I.C., shows seventeen photographs of dinosaur and human tracks in the Paluxy River bed, Texas, and devotes four pages to the many examples of human tracks discovered in Carboniferous formations— supposed to be 300 million years old![2]) These are some of

67

the more striking evidences in a remarkable chapter which attacks and demolishes the presupposition upon which is based all historical geology and the theory of evolution. That presupposition is uniformitarianism.

At this point it will be well to remind ourselves of the use and abuse of the doctrine where we have met it before. In one sense all science depends upon it, and no one denies its validity in explaining things as they are today. For example, Halley was a uniformitarian when he believed that a comet which had appeared at intervals of seventy-five and one-half years was probably the *same* comet, and would reappear after another seventy-five and one-half years. He was proved right by the event: the comet has a *uniform* cycle. This too is exactly what we would expect from Genesis 1:14, where God says that the sun, moon, and stars are for signs, seasons, days, and years. Regularity of motion and appearance is the outstanding feature of the heavenly bodies.

On the other hand, we perceived that uniformitarianism breaks down completely in the face of miracles, which are *not* day-to-day (or century-to-century) occurrences, but occasional and special acts of God. It is therefore impossible to explain creation on the basis of uniformitarianism, because creation was unique and unrepeatable ("the heavens and the earth were *finished,* and all the host of them. And on the seventh day God *finished* his work which he had made" [Genesis 2:1-2]). It will now be shown that uniformitarianism also fails to explain the phenomena connected with the Flood, because this event too has been declared by God to be unique and unrepeatable: "Neither will I again smite any more every thing living, as I have done" (Genesis 8:21*b*).

68

The Inadequacy of Uniformitarianism

This doctrine has been well explained by R. W. Fairbridge, Professor of Geology at Columbia University:

> In their effort to establish natural causes for the grand-scale workings of nature, 19th century geologists spurned the Scriptural concept of catastrophe. Under the leadership of the Scottish pioneers, James Hutton and Charles Lyell, they advanced the principles of uniformitarianism, which held that *the events of the past could be explained in the light of processes at work in the present.* [3]

But *The Genesis Flood* points out that this principle is quite inadequate to explain the following:

1. Volcanism and igneous rocks

Millions of square miles of these, thousands of feet deep, are found all over the world (e.g., the Deccan Plateau of India). No volcano ever known to man has produced anything like this quantity of lava, so it must have been produced by a process *not* repeated and *not* observed today.

2. Mountains

"All the major mountain ranges evidently were uplifted within the most recent eras of geologic history," yet "geologists are still unable to agree on a satisfactory hypothesis of mountain-building." Why is the earth not perfectly smooth all around? Why are mountains not coming up every day? Nobody knows. The past *cannot* be explained in the light of processes at work in the present.

3. Continental ice sheets

According to some geologists, four million square miles of North America and two million square miles of Europe were once glaciated. At least 29 "explanations" of this have been put forward, but every one has been found untenable in the light of further information. And we do not observe continental ice sheets forming today.

4. SEDIMENTATION

The argument here is highly technical, but we quote one example: most of Utah and Arizona, with large segments of Colorado and New Mexico—in all, some 250,000 square miles of plateau—have been uplifted from far below sea level (since most of its sediments are of marine origin) to over a mile above sea level, without at all disturbing the flatness of the strata! Such an uplift is quite unparalleled in historic times, and quite inexplicable in terms of uniformity. "By far the most reasonable way of accounting for the Grand Canyon is in terms of rapid deposition out of the sediment-laden water of the Flood."[4]

5. FOSSIL GRAVEYARDS

How and why were the five million mammoths of Siberia frozen to death in solid ice? How is it that in the water-laid "bone-bed" at Agate Springs, Nebraska, fossils are found of the rhinoceros, camel, giant boar, and numerous other animals not indigenous to America? How is it that in the Baltic amber deposits modern insects are found belonging to all regions of the earth? How could 800,000 million skeletons of vertebrate animals be entombed together in the Karroo formation (South Africa)? To account for these phenomenal mass burials, and many more like them, in terms of present-day processes is absolutely impossible.

6. COAL BEDS

"Regardless of the exact manner in which coal was formed, it is quite certain that there is nothing corresponding to it taking place in the world today."[5]

7. Footprints of extinct animals

(A photograph of dinosaur tracks is shown.) "It seems clear that the only way in which such prints could be preserved as fossils is by means of some chemical action permitting rapid lithification and some aqueous action permitting rapid burial. Sudden and catastrophic action is necessary for any reasonable explanation of the phenomena."[6]

8. Living fossils

Palaeontology is quite unable to account for the survival of some creatures which were supposed to have become extinct aeons ago. Among these the Tuatara, a New Zealand reptile, has no ancestors after 135 million B.C.; the coelecanth, a deep-sea fish, has no ancestors after 70 million B.C., and a mollusk named "neopilina galathea" has no ancestors after 280 million B.C.! How then did they leap the huge gap in the rocks? (If we believe the Bible, all is explained; some of their ancestors were fossilized in the Flood, but some survived. The Tuatara's ancestors presumably got a seat in the ark.)

9. Rock formations out of sequence

This evidence of itself is fatal to the theory of fossils being deposited at a uniform slow rate all over the earth, by "ages." This theory was framed by men who presumed (without proof) that any rock in which a "lower" creature is found must have been physically lower than any rock in which a "higher" creature is found; but the theory is contradicted by the facts over thousands of miles of the earth's surface. The "older" rocks are on top, and the "younger" rocks are below; and to account for this the "older" rocks are supposed to have been pushed into place by some enormous (but inexplicable) "overthrust"

71

from miles away. In *The Genesis Flood* many impressive examples are given, for example, in the Mythen Peak of the Alps, Eocene rocks (supposedly 60 million years old) are found under Triassic (supposedly 200 million years old). To account for this the Triassic rock, plus the Jurassic and Cretaceous above it, is said to have been pushed all the way from Africa to Switzerland!

CONCLUSION

I conclude this chapter with a quotation from Dr. E. M. Spieker, Professor of Geology at Ohio State University

> Does our time scale, then, partake of natural law? No ... I wonder how many of us realize that the time scale was frozen in essentially its present form by 1840? How much world geology was known in 1840? A bit of western Europe, none too well, and a lesser fringe of eastern North America. All of Asia, Africa, South America, and most of North America, were virtually unknown. How dared the pioneers assume that their scale would fit the rocks in these vast areas, by far the major part of the world? Only in dogmatic assumption ... and in many parts of the world notably in India and South America, it does not fit. But even there it is applied! The followers of the founding fathers went forth across the earth and in Procrustean fashion made the scale fit the sections they found, even in places where the actual evidence proclaimed denial. So flexible and accommodating are the "facts" of geology

72

13

CATASTROPHE COVERS IT ALL

When you have eliminated the impossible, whatever remains, however improbable, *must be the truth*.

SHERLOCK HOLMES

GEOLOGICAL EFFECTS OF THE FLOOD

All that now remains is to show positively that the Flood described in the Bible could have and would have been sufficient to cause all the geological effects which we have mentioned. (In *The Genesis Flood* many effects are listed which we have *not* mentioned; most of them will be of interest to specialists only.)

1. VOLCANISM, IGNEOUS ROCKS, AND MOUNTAINS

In Genesis 7:11 we read, "All the fountains of the great deep [were] broken up"; this would probably include vast submarine eruptions, and the uplifting of the world's major mountain ranges. "Oozing lava built great plateaus which now cover 200,000 square miles in Washington, Oregon, Idaho, and northern California. An even larger eruption created India's famous Deccan Plateau, whose once molten rock extends two miles below the surface."[1] The discovery of fossil fish high up in the Alps and other ranges strongly suggests that they were uplifted at this time.

2. Continental ice sheets

If these did in fact occur, they may have been formed *(a)* by the sudden change of weather caused by precipitation of the water canopy (which had acted as a "greenhouse," keeping the earth warm), or *(b)* by the upheaval of mountain chains with their snowcaps, or *(c)* by the great accumulation of ice near the poles.

3. Coal beds

As everyone knows, these were formed by enormous quantities of dead vegetation being subjected to heavy pressure. It is not so well known that the coal did not sink down "in situ," but was *water*-laid. Dr. Heribert-Nilson writes that only a process of immense magnitude and worldwide effect could account for the coal seams. More recently, in 1972, Dr. George R. Hill of the College of Mines and Mineral Industries, University of Utah, has demonstrated the rapid formation of coal from wood or other cellulosic material. He writes: "These observations suggest that in their formation high rank coals were probably subjected to high temperature at some stage in their history. A possible mechanism . . . could have been a short time, rapid heating event."[2]

4. Fossils

a. The Siberian mammoths may well have been overwhelmed when the "canopy" was precipitated and the resulting floods of water froze at the poles. The suggestion that the earth's axis was tilted out of the perpendicular at this time was first made by Edmund Halley and would seem quite credible since it is *after* the Flood that cold and heat and summer and winter are first mentioned (Genesis 8:22).

b. The animal graveyards are what one would expect in

a universal deluge: millions of fish were smothered by mud, and mammals of all kinds huddled together in caves to escape the rising waters. The disappearance of the rhinoceros from America is easily accounted for by the Flood, but hard to explain otherwise.

c. The *general order* of deposition of the fossils, too, is easily accounted for: at the bottom would be shellfish (the heaviest), then vertebrate fish, then amphibians, land reptiles, birds, and mammals. The larger and more mobile animals would have been able to keep out of the Flood longer than the smaller and less mobile. But on the other hand the currents must have ebbed and flowed with tremendous force, scouring, uprooting, overturning, and returning many times before they eventually subsided. This explains the frequent *dis*order of fossils in the sedimentary rocks; sometimes the dead trilobites got swept on top of the dead pterodactyls.

d. *Dinosaurs*. Many people are puzzled by the existence of these creatures which do not appear to be mentioned in the Bible. Why were they created? Were they represented in the ark? Why are they now extinct? Our answers can be only tentative.

It is possible that they are mentioned in Genesis 1:21 ("great sea-monsters"), though this would seem to include only the amphibians. Of course there is nothing to exclude them from "the beasts of the earth" (Genesis 1:25). They might have been herbivorous before the curse, or they might have been sent as a punishment for man's sin of violence (Genesis 6:11), as God threatened to do to His people in later history (Deuteronomy 32:24, KJV, "I will also send the teeth of beasts upon them." These beasts would be ordinary mammals, presumably). There may have been young dinosaurs aboard the ark, and these may

75

have been killed off by the postdiluvial climate. That they were once contemporary with man is suggested by (besides the Texas footprints) the ancient and widespread belief in *dragons*. Even the *Encyclopaedia Britannica* (14th ed.) allows that primitive peoples may have derived the idea from dinosaur bones, but it seems more probable that Noah's sons had seen live dinosaurs and passed them on to their posterity in the form of pictures.

For the present argument it is sufficient to know that the dinosaurs whose fossil remains we possess almost certainly died by *drowning*. The Dinosaur National Monument in Utah and Colorado is one of several huge graveyards found in various parts of the world. One writer describes it in these terms: "A majority of the remains of 300 dinosaurs probably floated down an eastward-flowing river until they were stranded on a sandbar. Perhaps the stegosaurs drowned trying to ford a tributary stream, or were washed down during floods." *The Genesis Flood* comments: "One could hardly ask for a better description of the way in which these great reptiles were overwhelmed, drowned and buried by the Deluge waters."

5. SEDIMENTATION

It is generally admitted that almost all the sedimentary rocks of the earth, which are the ones containing fossils, have been laid down by moving waters. The Oxford English Dictionary defines sediment as "earth or detrital matter deposited by aqueous agency." Obviously these great masses of sediment must have been first eroded, then transported, then deposited, "exactly the sort of thing that occurs in any flood, and which must have occurred on a uniquely grand scale during the great Flood of Genesis."

76

6. SUBMARINE GEOLOGY

Submarine geology confirms the concept of a universal Flood. Since 1950 large numbers of "drowned islands" have been discovered, many of them more than 6,000 feet below the surface, yet bearing abundant evidence that they were once above it! Also submarine canyons, hundreds of miles long, are very difficult to explain in terms of the present ocean level. Dr. K. K. Landes, chairman of the Department of Geology at the University of Michigan, writes: "Can we, as seekers after truth, shut our eyes any longer to the fact that large areas of sea floor have sunk vertical distances measured in miles?"[3] Uniformitarianism cannot account for this; but on the hypothesis of a Deluge precipated from the upper atmosphere, all is explained. Before the Flood there was much less sea in proportion to land, and the land was much flatter. When the rain fell, first of all the earth was covered to a great depth; then volcanism and "tectonic" activity built up the mountains; simultaneously the deep ocean beds sank down to accommodate the vastly increased waters on the surface of the globe. Thus the islands were "drowned," and the river canyons became submarine canyons. It is probably to this event that the psalmist in Psalm 104:6-8 refers:

The waters stood above the mountains.
At thy rebuke they fled;
At the voice of thy thunder they hasted away
(*The mountains rose, the valleys sank down*)
Unto the place which thou hadst founded for them.

7. THE ANTEDILUVIAN CLIMATE

Finally, the *antediluvian climate* powerfully confirms the concept of a universal Flood. In spite of the millions of

77

years supposedly separating trilobites from dinosaurs, and dinosaurs from mammals, it is firmly established that the climate of nearly *all* "ages" was warm and mild over the whole earth. For instance "in the Eocene age (60 million B.C.) subtropical heat was experienced in Greenland."[4] Uniformitarian geologists are at a loss to explain this tremendous difference from things as we see them today; they are forced to conclude that it must be due to changes in solar radiation. But Professor Hoyle states: "There is no evidence that changes take place in the radiation of the sun."[5]

On the other hand, one of the most original and convincing arguments of *The Genesis Flood* concerns the "canopy" already mentioned, for which the Bible phrase is "the waters which were above the firmament" (Genesis 1:7). They were invisible, held in suspension in the form of water vapor, and *they* caused the universally warm climate by trapping the sun's rays like a greenhouse. This accounts for the absence of rain before the Flood (Genesis 2:5), and the absence of clouds and rainbows too (Genesis 9:13 seems to imply this). Above all, it accounts for the Deluge; at God's command the vapor precipitated millions of tons of water, which first overflowed the earth, then drained off into the newly formed ocean beds. Only *after* this did the "normal" hydrologic cycle—evaporation, clouds, and rain—begin. This seems to be a thoroughly satisfying solution to the problem.

14

ARCHAEOLOGICAL DATING

It is a capital mistake to theorize before one has data.
SHERLOCK HOLMES

The "vapor canopy" seems also to be the answer to another question that is often asked: Has not Carbon 14 dating proved that the history of man goes back much farther than 4000 B.C.?

The Genesis Flood thoroughly discusses this and other chemical dating methods,[1] and the conclusion is that they all presuppose the *uniformity of the present atmosphere with the past*. "Carbon 14 dating assumes that the amount of C^{14} in the air has remained constant throughout the ages, but there is no proof, independent of the method itself, that cosmic-ray intensity has remained constant." Similarly, Professor Wilder Smith expounds at length the "built-in potential errors" of C^{14} dating[2]; and Professor W. F. Libby of California, who invented C^{14} dating in 1946, has admitted: "It is noteworthy that the earliest astronomical fix is at 4000 years ago, that all older dates have errors, and that these are more or less cumulative with time before 4000 years ago."[3] *The National Geographic Society News Bulletin* carried this statement:

79

"Analyses of iron in ancient bricks indicate that the magnetism may have declined by about two-thirds over the past 2000 years . . . Carbon 14 results from the collision of cosmic rays with nitrogen atoms in the air. If the amount of Carbon 14 has varied due to changes in the magnetic field, and has not remained a reliable constant for measuring age, *many estimates may be in error.*"[4]

The Genesis Flood mentions another very interesting fact: that the oldest trees in the world, the giant sequoias and bristlecone pines of California, are apparently of the *first* generation, since no one has discovered any old stumps of dead trees. Some of these trees are more than 4,000 years old; but the question is, Why are there none older since they appear to be immune to pests and disease? A probable answer is that their ancestors were wiped out by the Flood forty to forty-five centuries ago.

CAN WE TRUST THE TEXTBOOKS?

Christians who read books on archaeology would be well advised to take all the dates prior to 2000 B.C. with a large pinch of salt. Nearly all modern archaeologists entirely discount the Flood, and base their figures on one or more of these: (1) Carbon 14 dating, (2) layers of "culture" dug up on ancient sites, (3) ancient records of doubtful accuracy, (4) uniformitarian presuppositions. That all of these are very uncertain and admit of no truly scientific proof can be seen from the following examples:

1. "Dr. Stuart Piggott, a British archaeologist, reports that two radio-carbon tests on a sample of charcoal indicated a date of 2620-2630 B.C. for an ancient structure at Durrington Walls in England. But absolutely compelling archaeological evidence called for a date approximately 1000 years later."[5]

2. In "the world's oldest city," the Jericho Tower, now dated 6850 B.C., was dated 4800 B.C. in 1955 when Dr. Ehrich's book *Relative Chronologies in Old World Archaeology* was published. In the 1929 edition of the *Encyclopaedia Britannica* the date of the great pyramid was 4800 B.C. When I was in school in 1938 I was assured that the correct date was 3800 B.C. Today the date has dropped to 2850 B.C. These varying opinions would seem to indicate that archaeological dating is still somewhat less than an exact science.

3. The dates of the kings of Egypt are based chiefly on the writings of Manetho, a Greek priest (c. 285 B.C.). But one version of his list presents us with 561 kings who reigned 5,524 years, whereas another version lists 361 kings who reigned 4,480 or 4,780 years. Further, many of the dynasties may have reigned at the same time in different parts of Egypt. Alexander Hislop wrote:

> Bunsen casts overboard all Scripture chronology and sets up the unsupported dynasties of Manetho as if they were sufficient to override the Divine word as to a question of historical fact. But if the Scriptures are not historically true, we can have no assurance of their truth at all. It is worthy of note that though Herodotus vouches for the fact that there were no fewer than twelve contemporaneous kings in Egypt, Manetho has made no allusion to this, but has made his Thinite, Memphite, and Diospolitan dynasties of kings, and a long etcetera of other dynasties, all successive![6]

Manetho also lists the "reign of the Gods" as lasting 13,900 years. That his work has some value is not denied, but it cannot be compared with the Bible for trustworthiness. Martin Anstey writes: "The Egyptians themselves

81

never had any chronology at all. They were devoid of the chronological idea."[7]

4. Dr. Paul Thieme, Professor of Sanskrit and Comparative Philology at Yale University, writes as follows:

> Indo-European, I conjecture, was spoken on the Baltic coast of Germany late in the fourth millennium B.C. (i.e. about 3200). Since our oldest documents of Indo-European daughter languages (in Asia Minor and India) date from the second millennium B.C., the end of the fourth millennium would be a likely time anyhow. 1000 or 1500 years are a time sufficiently long for the development of the changes that distinguish our oldest Sanskrit speech from what we construct as Indo-European.[8]

But this theory allows no place for the Tower of Babel and the confusion of tongues, which (according to Moses, Genesis 11:7) was effected instantaneously and did *not* evolve over many centuries. (We do not, of course, deny that there has been development *since* the miracle at Babel.) Like Professor Gamow's speculation as to the origin of the universe and solar system, Professor Thieme's "conjecture" about the origin of Sanskrit is squarely based on uniformitarian presuppositions unsupported by any historical evidence whatever.

15

THE ORIGIN OF LANGUAGE

What they painfully reconstruct from a million dots, arranged in an agonizing complexity, He [God as Artist] really produced with a single lightning-quick turn of the wrist . . . His mind obeying laws of composition which the observers, counting their dots, have not yet come within sight of, and perhaps never will.

C. S. LEWIS, *Miracles*

The phenomenon of language appears to present an impassable stumbling block to the theory of man's evolution. In their first flush of enthusiasm some Darwinians attempted "to prove from the anatomical structure of the skulls of the earliest prehistoric men that they could not have possessed the faculty of speech . . . but this conclusion is certainly drawn from insufficient premises, and has no foundation in fact."[1] Assuming that the first men could *not* talk, evolutionists put forward various theories of the origin of language. The best known of these are the bow-wow, pooh-pooh, and yo-he-ho theories. Men are supposed to have imitated animals sounds (bow-wow) or uttered instinctive emotional cries (pooh-pooh) or natural singsong at teamwork (yo-he-ho: the Volga boatmen).

This, according to nineteenth-century experts, was the beginning of speech. Obviously, then, the more primitive a language, the more monosyllabic it should be.

But modern research among the "primitive" peoples of the world has proved these ideas to be pure fantasy. Dr. Eugene Nida writes:

> During the last century there was a popular evolutionary belief that languages had evolved from monosyllabic structure (such as Chinese) through the agglutinative structure of such languages as Bantu and Aztec to "the highest form of speech", the inflected languages of Europe. Such blatant egoism has been found lacking any basis in fact. Even Chinese, which was cited as such a primitive language, was discovered to have had some inflection in its earlier history. As for "primitive languages", they have been shown to exhibit all the types of structure found in any language spoken by "civilized" peoples . . . languages are arbitrary systems: there is nothing in the nature of the sounds themselves which makes it obligatory for them to carry particular meanings . . . even exclamations show no basic similarities. We yell "Ouch!", but a Spanish-speaking person cries "Ay! Ay!" Our dogs bark bow-wow, but the Kipsigis of Kenya insist that dogs say 'u 'u. It is entirely arbitrary as to which sounds are to be employed to represent particular ideas or emotional responses.[2]

Earlier Dr. Nida states that "there is no tribe of people anywhere in the world which does not have thousands of words in its vocabulary, and an intricate systematic way of putting words together into phrases and sentences, i.e. a grammar." The Yaagans of Tierra del Fuego—a nomadic tribe—have a vocabulary of 30,000 words, as do the Zulus of South Africa. "Almost any verb root in Aymara (Peru) can occur in at least 100,000 different combinations."

Some Bantu languages have a grammar more rigid and precise than Greek: "each word must come in a specific order and begin with a prefix which indicates the system of modification . . . Wintu Indians of California have special forms which indicate whether a statement is (1) hearsay (2) a result of direct observation, or (3) inferred, with three degrees of plausibility."

Among the great scholars of the Victorian era, few if any excelled Richard Chevenix Trench as a philologist. From a study of worldwide missions he concluded that language is not an art (like toolmaking) but an instinct or God-given faculty, like the bee's instinct to make cells and the bird's to make its nest. Why? Because there are tribes which use no tools and cannot even cook, but "there have never yet been found human beings who do not employ speech." Second, he maintains that "the theory that the savage was the seed out of which in due time civilized man was unfolded . . . is contradicted by every notice of our actual experience. Here, as in everything else that concerns the great original institutes of humanity, our best and truest lights are to be gotten from a study of the first three chapters of Genesis. What does the language of primitive savages on close inspection prove to be? In every case it is the remnant and the ruin of a better and nobler past . . . as one habit of civilization after another has been let go, the words which those habits demanded have dropped as well, first out of use, then out of memory." The Bechuanas of South Africa at one time used the word *Morimo* to mean "Him that is in Heaven," a supreme Divine Being. But Moffat in 1840 found that the word and the idea had almost vanished from the Bechuana vocabulary of his day. "Here and there he could meet with an old man, scarcely one or two in a thousand, who remembered in his youth to have heard speak of 'Morimo.'"[3] This example shows that there is nothing "automatic" about a language improving or expanding. When the Bantu invaded South Africa they may have had a much richer vocabulary than at present.

ANCIENT LANGUAGES

What about ancient languages? If evolution be true, will not the oldest be the simplest?

Everyone knows that Latin is much harder than English—cases, genders, moods, voices, personal terminations, and precise syntax. Greek, perhaps 600 years older than Latin, is still more difficult; and when we come to Vedic Sanskrit (c 1500 B.C.) the complexity is almost unbelievable.

> In the Vedic language the verbal system is of considerable complexity. A verb might have various stems, viz. present (sometimes more than one), aorist (three), perfect (characterized by reduplication and peculiar terminations), future. The various present stems indicated various types of present-stem action, such as intensive, repetitive, inchoative, causative, desiderative etc. Each of the first three stems had five moods—indicative expressing fact, injunctive and subjunctive expressing will and futurity, optative and imperative. In the indicative of the present, perfect and future stems there were two tenses, present and past . . . each tense had three persons and three numbers—singular, dual and plural. Finally, each tense could be conjugated in two voices with different terminations—active and middle. Among the parts of the infinitive verb there was connected with each stem a participle which could be either active or middle, and independent of tense stems a past principle, one or more infinitives, a gerundive and an indeclinable participle or gerund. The total number of possible forms belonging to any one

verb is thus very great . . . this verbal system was *greatly simplified in Classical Sanskrit* (500 B.C.—1000 A.D.).[4]

MODERN ENGLISH

Coming down to modern English, in which most verbs have less than a dozen possible forms (e.g., do, dost, does, did, didst, done, doing), we realize that we are at the end of a long line of *devolution*. So far from being the highly complex descendant of a simple nongrammatical ancestor, our speech is the simple (comparatively) monosyllabic (comparatively) nongrammatical descendant of a highly inflected, complex, polysyllabic, and exactly grammatical ancestor!

INDIAN PARALLELS

Precisely the same phenomenon can be observed in India. The most direct descendant of Sanskrit, Hindi, is only some 400 years old and is generally reckoned to be the easiest to learn of all Indian languages. Tamil, probably at least 1,000 years older than Hindi, is accounted the hardest. It is highly inflected, having eight cases for every noun, a positive and negative verb, and a different termination for every person and every sex (masculine, feminine, neuter) in every tense of the verb. Malayalam was originally a dialect of Tamil; now it has discarded all personal terminations in the verb, and the person is indicated by the pronoun only (cf. our "I did, you did, he did, we did, they did").

We conclude, then, that in language the natural direction is *down,* from the higher and more difficult to the lower and easier. Nowhere on the face of the earth is there evidence of evolution from simple to complex; everywhere the evidence points to devolution from complex to simple.

87

A second very interesting fact discovered by modern research is that there are at least fifty distinct families of languages. Nine of them include nearly 90 percent of the world's population—Indo-European, Sino-Tibetan, Semitic-Hamitic, Dravidian, Ural-Altaic, Japanese, Malayo-Polynesian, Bantu, and Austroasiatic; and the approximately remaining forty are spoken by comparatively small groups, for example, the Basques of the Pyrenees, whose language cannot be related to any other in Europe.[5] Between these families there is no evidence of any common source or historical connection (although there are linguists who believe that such a connection will one day be proved, and still search for it). Japanese, for example, is totally different from Arabic, and both are totally different from Bantu. Yet, almost all anthropologists now admit the unity of the human race. Why then are our languages so distinct? If we rule out the theory of random evolution from monosyllables, for which there is no evidence whatever, there seems to be only one possible answer: the story of the Tower of Babel in Genesis 11 is literally and historically true.

Speech was God's gift to man at his creation. Adam was able to understand verbal commands, to name the animals, to name his wife (*not* at random, but meaningfully), to talk with her, and with God. Till a hundred years after the Flood, it seems, "the whole earth was of one language, and of one speech" (Genesis 11:1). Most commentators take this to mean that they used the same vocabulary and the same pronunciation. Then God did a miracle of judgment, instantly confounding their language so that they might not understand one another. Hence, Japanese, Arabic, Bantu, etc. It is true that an Englishman, a

Frenchman, and a German who knew not one word of each other's language would be equally at a loss to understand each other; but the actual evidence suggests that God not merely split up one language family into many related branches, but initiated many perfectly distinct methods of verbal communication. Every one of these languages, we may presume, was highly complex and included a large vocabulary. Over the course of centuries some fortunate tribes (like the Greeks) learned to write, and produced a brilliant literature. Others got lost in the jungle. But even the most "primitive" tribes still retain in their language a relic of the glorious past, a proof that they are cousins of those who built the pyramids, and of those who fought at Troy.

CONCLUSION

We conclude, then, that as astronomers have failed to explain the origin of the solar system in terms of cosmic evolution, as geologists have failed to explain the origin of mountains and "drowned" islands, as biologists have failed to explain the origin of species—let alone of man—so archaeologists, anthropologists, and linguists have completely failed to explain the origin of language in terms of naturalism, uniformitarianism, and evolution. Once again the Bible supplies us with the only explanation perfectly congruous with all the known *facts*.

16

WHERE HAS SCIENCE GONE WRONG?

The idols of the theatre are the authoritative opinions of others which a man likes to accept as a guide when interpreting something he hasn't experienced himself. . . .
Another idol of the theatre is our over-willingness to agree with the arguments of science. One can sum this up as the voluntary acceptance of other people's errors! "That's good," said Oleg . . . "Voluntary acceptance of other people's errors! That's it!"

ALEXANDER SOLZHENITSYN, *Cancer Ward*

Two questions remain to be answered:

1. If all these facts are so, why do most modern scientists reject the Bible account of the Flood and cling to the geological time scale? We believe the answer is found in the Bible itself:

"In the last days there will come men who . . . will say, 'Where is now the promise of his [Christ's] coming? Our fathers have been laid to their rest, but still everything continues exactly as it has always been since the world began." (Uniformitarianism!) "In taking this view they lose sight of [*deliberately ignore* or willfully forget] the fact that . . . by water that first world was destroyed, the water of the deluge" (2 Pe 3:3-6 NEB). God has set His unmistakable mark upon the earth. The destruction of billions of His own creatures, the burial of trillions of tons of vegetation, the depression of old continents and the uplifting of new ones — these mighty acts proclaim the power and holiness of our Creator. But the geologists, biologists, and archaeologists (like some astronomers) "glorified him not as God, neither were thankful; but became vain in their imaginations" (Romans 1:21, KJV). *This* accounts for their misconstruing the evidence and turning it against God's Word. Only in the last two centuries have scientists come to doubt the reality of the universal Flood, and Peter tells us this is a sign of the *last* days before the coming of Christ.

2. If all these facts are so, why do many sincere Christians believe that the universe is millions of years old, and that the Flood was only a local event?

Perhaps these sincere Christians have been "carried away with the error of the wicked," because the particular error against which Peter here warns us (2 Peter 3:16-17) is the misinterpretation of Scripture. Paul wrote to sincere Christians in Colosse about the dangers of "philosophy and vain deceit" (Colossians 2:8), and to sincere Christians in Ephesus about the "sleight of men . . . craftiness . . . error" (Ephesians 4:14). Church history shows that sincere Christians have often erred from the truth. There have been other periods when all Western civilization was the victim of a gigantic hoax, when millions of people were persuaded to abandon God's truth in the Bible and embrace "hollow and delusive speculations, based on traditions of man-made teaching" (Colossians 2:8, NEB).

91

GNOSTICISM

Our first example is the Gnostic heresy. When Paul warned Timothy against "oppositions of *science* falsely so called," he used the word "gnosis." Professor F. F. Bruce has written:

> The general name given to the new learning was *gnosis*. This is simply the Greek word for knowledge, but it tended to be used in a superior sense, much in the same way that more recently the Latin word for knowledge, *scientia* or science, has come to be spelt with a capital letter and used almost personally as the subject of sentences. "Science tells us" that such and such is the case; that was very much the way in which people of those days spoke of *gnosis*. When Christianity made headway in the Greek world, it soon came into collision with the possessors of *gnosis*, who were Gnostics (the people who possess real knowledge). The result was an attempt *to restate Christianity in terms of gnosis, to fit it into the current cosmology.*[1]

Now it would be not unfair to say that most commentaries on Genesis since 1870 are exactly this—attempts to restate the Bible doctrines of creation, the Fall, and the curse, in terms of the current cosmology—"the assured results of Science."

Professor Bruce continues:

> The new theosophy was very attractive, and throughout the second century it made considerable headway among the more intellectual Christians of the Graeco-Roman world. While it was pre-Christian in origin, deriving elements from pagan thought, and absorbing a good deal of sheer magic in the process, it developed a variety of definitely Christianized forms.[2]

And the *Encyclopaedia Britannica* states: "One of the determining forces of Gnosticism was a *fantastic oriental imagination.*"

We may confidently predict that in ages to come other historians will write somewhat as follows (with apologies to F. F. Bruce): "The new scientism was very attractive, and throughout the nineteenth and twentieth centuries it made considerable headway among the more intellectual Christians of the Western world. While it was pre-Christian in origin, deriving elements from pagan thought, it developed a variety of definitely Christianized forms (e.g., theistic evolution). However, it was finally exposed as nothing but another 'hollow and delusive speculation' of men who totally misconstrued the evidence deposited by the Noahic Deluge. Christians at last awoke to the fact that all the fossils could be explained in terms of the Flood, and that one of the determining forces of scientism was a *fantastic occidental imagination* which could explain every irregularity in the solar system without explanation, leap the gaps in the atomic series without evidence (by 'sheer magic'), postulate the discovery of fossils which have never been discovered, and prophesy the success of breeding experiments which have never succeeded. Of this kind of science it might truly be said that it was 'knowledge falsely so called'!"

MEDIEVAL SUPERSTITION

In their greed for money they will trade on your credulity with sheer fabrications.

2 PETER 2:3 (NEB)

Another astonishing example of credulity is the medieval belief in relics. Starting as veneration for the martyrs in the second and third centuries, this grew into a reverence for anything supposed to have had some connection with the saints or the Saviour. Frauds, pious and impious, flooded the market. At the cathedral of Trier the seamless robe of Christ was "discovered" early in the twelfth century; the bodies of the "three kings" (who saw the star) were deposited at Cologne in 1164 by Frederick I; at least two French churches possessed "the crown of thorns"; King Athelstan (A.D. 930) donated to the monastery at Exeter fragments of "the candle which the angel of the Lord lit in the tomb of Christ, of the Burning Bush, and of one of the stones which slew Stephen." In 1520 Frederick the Wise, Luther's protector, had 19,013 such relics in the Schlossekirche at Wittenberg. Thus thousands of sincere and intelligent Christians over hundreds of years were completely deceived, not only as to the spiritual efficacy of these relics, but also as to their *history*, which of course had no connection with the acts or facts of the Bible.

Let no one think that the human race is less gullible today than 700 or 1,700 years ago; the gullibility has changed not in degree but in direction. As C. S. Lewis wrote in *Screwtape Letters*, the devil uses fashions in thought to distract men from their real dangers. In an age of magic and "miracles," faith can quickly degenerate into superstition; and in a skeptical antisuperstitious age men can easily be persuaded that "miracles don't happen," that only science is infallible, that whatever the "experts" say must be true, even when there is no evidence whatever to support their statements. Today these unproven statements are held in the same naive veneration as were the relics of the Middle Ages.

Modern Scientism

When men cease to believe in God, they do not believe in nothing, they believe in anything.

G. K. Chesterton

Two final examples will show the cul-de-sac into which evolution has led its devotees.

1. As everyone knows, some marine creatures are not fish but mammals; over one hundred species of whales (including dolphins and porpoises); the dugong of the Indian Ocean, and the manatee of tropical American and African waters; and forty-seven species of seals (including sea lions, fur seals, true seals, sea elephants, and walruses) numbering about twenty-five million—"the largest surviving group of big carnivorous animals in the world today."[3] Evolutionists confidently affirm that all these were once land animals which one day took a walk into the sea (their fishy ancestors having crawled *out* of the sea in order to become land animals) and by a process of gradual adaptation were transformed into the highly efficient swimmers and divers which we now see. A land mammal could not, of course, become a whale overnight. It is estimated that at least thirty intermediate forms would have been necessary. But what is *not* stated in the evolutionary textbooks is this: that among all the thousands of fossils which have been examined, *not one has ever been discovered* which could begin to bridge the gap; not one half whale, half walrus, or half seal; not one "link" which might be part of the chain connecting land animals with "the largest group of big carnivorous animals in the world today"! And it was T. H. Huxley who wrote:

"The primary and direct evidence in favour of evolution can be furnished *only* by palaeontology [fossil remains]...

95

if evolution has taken place, there will its mark be left: if it has not taken place, there will be its refutation."

2. The same can be said of another mammal which apparently got bored with its natural environment and found no difficulty in constructing for itself the most efficient flying machine ever invented, plus built-in radar. We refer, of course, to the bat, whose marvelous dexterity in flight has been proved to excel any bird's. There are several hundred species of them, inhabiting most parts of the globe, and numbering many millions. Once again the evolutionist is forced to postulate intermediate stages for the development of a ground mammal into an air mammal; in this case they reckon the figure is twenty. But once again the links are missing; not a single fossil has ever been discovered of *any* creature which can be called a "half bat," let alone twenty distinct stages. Science has accepted the dogma of the bat's evolution in spite of a *total lack of paleontological evidence.*

If we pause a moment to inquire whether there is any connection between the medieval belief in relics and the modern belief in evolution, we shall find that both delusions stem from a common principle of *transferred authority.* People accepted the authenticity and efficacy of relics because the Church which guaranteed them was founded upon truth: the life, death, and resurrection of Jesus Christ, and the New Testament. These truths commended themselves to the conscience of every honest man; and because the custodians of these truths ordered the worship of relics, many honest people felt that they should obey. The authority which the Church had rightly used to preserve and propagate the historic faith was wrongly used to preserve and propagate unhistorical fantasy and fiction.

Christendom swallowed a lie because it was served up by the purveyors of truth.

Today the greatest authority in the world is the authority of science. Science has put men on the moon and new hearts into old bodies. It is science that we bless for our creation, preservation, and all the blessings of this life, but above all for the inestimable boon of television, automation, and contraception. What the Church's authority was to medieval Europe, that the authority of science is to Western man. And just as Frederick the Wise accepted papal authority in matters reaching far beyond the pope's rightful jurisdiction, so modern man has accepted the dogmas of science regarding the origin and age of the universe, even though these are subjects quite outside the legitimate sphere of scientific investigation. The respect due to scientists for their undoubted achievements in the realm of the seen, the tangible, and the predictable, has been naively transferred to their unwarranted pronouncements concerning the unseen, the intangible, and the unpredictable. The fallacy of evolution has been swallowed because it is prescribed by many searchers after truth.

We may leave it to the historians of a later age to decide which creed requires the greatest credulity: the Gnostic belief of early centuries, the medieval faith in relics, or the modern belief in "missing links"—a belief which bridges enormous gaps in the fossil record with creatures of the imagination.

What is certain is that civilization has once again been the victim of a gigantic hoax; in almost every university of the world a stupendously improbable nonfact is being taught as if it were true.

17

EPICURUS RESURRECTED

About 306 B.C. a man named Epicurus established a school of philosophy at Athens. Disgusted with the popular religion of his day—legends of cruel and capricious gods—he framed a system which excluded all divine interference from the universe. According to Epicurus, the infinite variety of nature is explained by innumerable atoms perpetually falling through space; each atom has a "swerve" or bias which makes it collide with other atoms by chance, thus producing birds, beasts, trees, bees, stars, and everything else—"by chance." Epicurus was a transformist. He taught evolution guesswork.

Two centuries later a Roman poet, Lucretius, embraced this philosophy, hailed Epicurus as the savior of mankind, and with apostolic zeal flung Epicurean physics and ethics into Latin verse to convert his fellow countrymen. How flattered Lucretius would have been to learn that 2,000 years later he would be quoted in a biology textbook for his "science"! Professor Sir Gavin de Beer writes approvingly: "Chance was exactly what Lucretius invoked . . . to explain living organisms."[1] But "chance" is only a word invented by humans to conceal our ignorance. It explains nothing. If we perfectly understood all the laws of motion, we could infallibly predict whether a coin will come down heads or tails. A Christian believes that God *does* perfectly understand His own laws and knows which side up the coin will land, but Epicureans and neo-Darwinians believe that *nobody* knows!

> All things bright and beautiful,
> All creatures great and small,
> All things wise and wonderful—
> The Lord *Chance* made them all!

Do we want this taught to our children?

What a tragedy that while Charles Darwin, the apostle of chance, is universally famous, and thousands of children are taken every year to worship at his shrine in South Kensington, England, the names of really brilliant scientists like Faraday, Clerk Maxwell, and William Thomson are comparatively little known. Yet *their* theories and inventions *really work,* and all three were firm believers in the Bible. Clerk Maxwell, founder of the Cavendish laboratory at Cambridge, "showed on fit occasion his contempt for that pseudo-science which seeks for the applause of the ignorant by professing to reduce the whole system of the universe to a fortuitous sequence of uncaused events."[2]

Neo-Darwinism in its basic philosophy of the universe is simply Epicureanism revived, and the head-on clash with Christianity is as inevitable as when Paul met Epicureans in the forum at Athens and declared: "God who made the world and all things therein . . . is Lord of heaven and earth." No one would object if Epicureanism/Neo-Darwinism were studied at the university level in the department of philosophy. But the scandal is that in nearly

99

every *school* in America and Britain, children are being indoctrinated with this ancient Greek philosophy masquerading as "modern science"!

18

BACK TO THE BIBLE

I prefaced this book with a lighthearted poem because the British people have been endowed with a strong vein of humor and common sense. Isn't it time we applied these excellent qualities to education?

The desperate search for pedigrees of the whale, the bat, the giraffe, and a thousand other creatures is worthy of Carroll's *Through the Looking Glass*. The frequent discoveries of various shades and grades of homo sapiens or insipiens, anthropopithecoi or pithecanthropoi, each discovery exploding the fragile hypotheses erected upon the last discovery, are reminiscent of Gilbertian comedy. For entertainment and delight let us send our children to Kipling. *How the Elephant Got His Trunk, How the Camel Got His Hump, How the Rhinoceros Got His Skin,* and *How the Leopard Got His Spots* are tales as plausible as any concocted by the solemn theorists. But for serious history and prehistory let us turn again to the Bible, for there alone will be found the *facts* which explain our wonder-filled world.

Let professors revert, if they will, to the automated hit-or-miss machine of Epicurus; but children will be wiser to heed the words of Britain's "greatest Englishman," Sir Winston Churchill:

> We believe that the most scientific view, the most up-to-date and rationalistic conception, will find its fullest satisfaction in taking the Bible story literally . . . we may be sure that all these things happened just as they are set out in Holy Writ. We may believe that they happened to people not so very different from ourselves, and that the impressions those people received were faithfully recorded, and have been transmitted across the centuries with far more accuracy than many of the telegraphed accounts we read of today's events.
>
> Let the men of science and of learning expand their knowledge and probe with their researches every detail of the records which have been preserved to us from those dim ages. All they will do is to fortify the grand simplicity and essential accuracy of the recorded truths which have lighted so far the pilgrimage of man.[1]

EPILOGUE

Two million monetary pounds of the notes lost in Britain's great train robbery may never be recovered. But spiritual and intellectual riches, though lost to one generation, may be recovered in the next.

Even now it is not too late to restore to our children their rightful heritage—the whole Bible as the only sure basis for education in religion, in history, *and in science*.

Or, at the very least, children should be informed that there are *two* conflicting views on origins: *some* scientists believe in evolution, others in special creation. The arguments on *both* sides should be presented, and children should be *free to choose* between them.

The present dogmatic teaching of evolution guesswork as "fact" is closely akin to brainwashing, and is indefensible on any principle of logic, ethics, or democracy.

Appendix A

THE SIGNIFICANCE OF "LET THERE BE LIGHT"

1. "As is now clear, light is only one form of energy. Most probably all forms of energy were called into use or being during this first creation day. The various stars and galaxies were then created by conversion of energy into mass according to the formula

$$\frac{e\ (energy)}{c^2\ (speed\ of\ light\ squared)} = m\ (mass)$$

. . . the whole universe became one vast system of light and energy, since one cannot from the viewpoint of physics conceive of visible light as distinct from other forms of energy. By the fourth day the conversion of energy into mass evidently reached a concentration in the various systems of atomic furnaces which we now recognize as the sun and stars. It should be emphasized that, however vast the universe may be, *light photons from the most distant stars would be visible immediately,* since the stars were made by conversion of light into closed orbits of energy, which we call mass.

"A crude analogy is that of filling a large tank with water under pressure through a hose several hundred feet long. Once the tank is full the flow immediately reverses when pressure is discontinued. No matter how long the hose, water pours out *immediately* at a rate determined by the tank pressure. Astronomers of the uniformitarian school would have us starting with an empty hose. Then, of course, the time taken by the water to travel through the hose would be a measure of the length of the hose. So they assume stars beginning to shine with no photons of light connecting them with the earth or other stars. But if the stars are conceived of as being created by the flow of energy into them, then as soon as they begin to shine by virtue of this accumulated energy, a reversal in flow of light photons would *immediately* be visible here on the earth.

2. "The creation account, by stating that the sun, moon and stars were not activated until the fourth day, indirectly supports the Copernican system of astronomy. For if, as Ptolemy assumed, the sun in its daily cycle around the earth caused Day and Night, how could there have been nights and days before the sun gave light? *The answer, of course, is that the earth's rotation gives our diurnal cycles,* and always has, since God said, 'Let there be light'. This light came directly from Him until the fourth day, by which time the sun was activated as suggested above."[1]

Appendix B
THE BIBLE AND SCIENCE

"Though the Bible is not a revelation of Science it may be expected to be free from error, and to contain, under reserved and simple language, much concealed wisdom, and turns of expression which harmonize with natural facts, known perfectly to God, but not known to those for whom at first the revelation was designed.

"The Bible presents a striking contrast to the sacred books of heathen nations. All systems of religion, and all eminent philosophers of antiquity, maintained notions of science no less absurd than their theology.

"In Greek and Latin philosophy the heavens were a solid vault over the earth, a sphere studded with stars, as Aristotle called them. The sages of Egypt held that the world was formed by the motion of air and the upward course of flame: Plato, that it was an intelligent being: Empedocles held that there were two suns: Zeucippus that the stars were kindled by their motions, and that they nourished the sun with their fires.

"All eastern nations believed that the heavenly bodies exercised a powerful influence over human affairs, often of a 'dis-astrous' kind, and that all nature was composed of four elements—fire, air, earth, and water, substances certainly not elementary.

"In the Hindu philosophy the globe is represented as flat and triangular, composed of seven storeys; the whole mass being sustained upon the heads of elephants who, when they shake themselves, cause earthquakes. Mohammed taught that the mountains were created to prevent the earth from moving, and to hold it as by anchors and chains.

"The Fathers of the Church themselves teach doctrines scarcely less absurd. 'The rotundity of the earth is a theory which no one is ignorant enough to believe,' says Lactantius.

"How instructive, that while every ancient system of idolatry may be overthrown by its false physics, not one of the 40 writers of the Bible, most of whom lived in the vicinity of nations that held these views, has written a single line that favours them. This silence furnishes a striking confirmation of the truth of their message."[2]

Appendix C
DEMYTHOLOGIZING GENESIS

In a brilliant exposé of Bultmann's school of New Testament criticism, C. S. Lewis writes: "All theology of the liberal type involves at some point—and often involves throughout—the claim that the real behaviour and purpose and teaching of Christ came very rapidly to be misunderstood and misrepresented by His followers, and has been recovered or exhumed only by modern scholars."[3] He goes on to show that very much the same claim was made by Jowett in reinterpreting Plato, and is made "every week by a clever undergraduate, every quarter when a dull American don discovers for the first time what some Shakesperian play really meant." Lewis concludes: "The idea that any man or writer should be opaque to those who lived in the same culture, spoke the same language, shared the same habitual imagery and unconscious assumptions, and yet be transparent to those who have none of these advantages, is in my opinion preposterous. There is an *a priori* improbability in it which almost no argument and no evidence could counterbalance."

The same argument may be applied with equal force to the Old Testament, and to Genesis 1-11 in particular. Modern scholars reinterpret Moses and claim to have discovered for the first time what these chapters really mean; but it is rather more probable that the true interpretation is to be found among those who lived in the same culture, spoke the same language, shared the same habitual imagery and unconscious assumptions—in other words, the Jews. And it is certain that the Jews regarded the "days" as literal days, the Flood as universal, and the genealogies as constituting a chronology.[4]

Second, C. S. Lewis points out that the question, Do Miracles happen? "is a purely philosophical question. Scholars, as scholars, speak on it with no more authority than anyone else . . . if one is speaking of authority, the united authority of all the biblical critics in the world counts here for nothing. On this they speak simply as men; men obviously influenced by, and perhaps insufficiently critical of, the spirit of the age they grew up in."

Similarly, on the question, Was creation an instantaneous miracle? the united authority of all the scientists in the world counts for nothing. Harlow Shapley, George Gamow, Fred Hoyle, Julian Huxley, and a host of others may believe it was *not;* but here scientists, as scientists, speak with no more authority than anyone else. On this they speak simply as men, men obviously influenced by, and perhaps insufficiently critical of, the spirit of the age they grew up in.

Third, it is instructive to note how unbelief runs true to form both in scientist and scholar. The New Testament critic, faced with a document which purports to record definite miracles and dogmatic statements, tries to explain them away by postulating hypothetical vanished documents (e.g., "Q") and hypothetical church situations from which the "mythical" stories of the Gospels are supposed to have "evolved." By writing book after book on the subject, with an immense display of historical erudition, the scholar succeeds in convincing many of his contemporaries that things actually did happen as he imagines they might have happened. His readers forget that the whole structure is a vast inverted pyramid of assumptions built upon a pinhead of fact. C. S. Lewis comments:

"I have watched with some care similar imaginary histories both of my own books and of books by friends whose real history I knew . . . my impression is that in the whole of my experience not one of these guesses has on any one point been right: the method shows a record of 100 per cent failure. 'The assured results of modern scholarship' as to the way in which an old book was written, are 'assured', we may conclude, only because the men who knew the facts are dead and can't blow the gaff."

In very much the same way a modern cosmogonist, faced with the miracle of a "going" universe, builds up an imaginary history of atoms, nebulae, stars, the solar system. From there the biologist takes over and "explains" the present order of living things in terms of evolution. Where facts do not exist, they can be assumed. But we may safely conclude that what is true of unbelieving scholarship is equally true of unbelieving scientism: the "assured results of modern science" as to the way in which this old world was made, are "assured" only because the men who knew the facts (Adam, Noah, etc.) are dead and can't blow the gaff. The reconstructed histories of the universe, of life, and of mankind, are (as Lewis says of essays on *Piers Plowman*) "most unlikely to be anything but sheer illusions." He that sitteth in the heavens shall laugh them to scorn, because from my point of view not one of these guesses (where they conflict with Genesis) has on any one point been right. The method shows a record of 100 percent failure.

110

Appendix D

ARE GENESIS 5 AND 11 INTENDED TO BE STRICT CHRONOLOGY?

In this matter Herodotus is but a mushroom . . . the Vulgate [i.e., the Bible] handles matters three thousand years before him whom pedants call the Father of History.

THE CLOISTER AND THE HEARTH

The Bible gives to our new world, and to this proud race of ours, so young an age, that men of all times : . . have foolishly revolted at it.

L. GAUSSEN

The explanation of Genesis 5 and 11 offered in *The New Bible Dictionary* is this: that since there are gaps in the genealogy of Christ in Matthew 1, therefore we may assume that there are similar gaps in the genealogies of those chapters. This theory is open to at least three objections:

1. A fundamental principle of Scripture is its complementarity. What one book omits, another includes; what one writer narrates in full, another summarizes. The Bible is meant to be read like any other book, from the beginning to the end; and, as in any other book, that which has been clearly stated in earlier "chapters" is assumed as known in later "chapters." Thus when Matthew, writing to the Jews, stated that "Uzziah begat Jotham" (Matthew

111

1-9), he was not misleading or deceiving anyone. Probably Jewish children knew their list of kings as well as an English schoolboy knows the list of *his;* and they would immediately have spotted that three names had been omitted. They would perceive that Matthew was not attempting to give a complete genealogy, but simply to record the Messiah's pedigree in a way that is easy to remember.

On the other hand, if the author of Genesis 5 and 11 has omitted names from the list, he has misled scores of generations of Jewish and Christian scholars, because there is no possible way of checking the record, and all have therefore assumed that the record is complete. Can we think that He who is the truth would lead so many thousands, for hundreds of years, to believe a lie?

2. In any discussion of Genesis 5 and 11 the relevant New Testament document is not, surely, Matthew's genealogy, but Luke's (chap. 3). And it seems very evident that Luke intends to give us a *complete* list of *all* the names. He gives forty-two names covering the 1,000 years from David to Jesus, less than twenty-five years per generation. Can we believe there have been any omissions here? And if there are no omissions in the first part of the genealogy, are we justified in postulating omissions in the second?

3. However, by far the strongest objection to the gap theory is this: it makes nonsense of Scripture. For whereas all the other genealogies in the Bible consist merely of lists of names, Genesis 5 and 11 give the age of each father at the birth of his son, thus absolutely excluding the possibility of any gap. The so-called parallels in Egyptian literature are not parallel at all, for none of them mention any age before or after the father's "begetting." And if we deny Moses' purpose to construct a complete

112

chronology from Adam to Joseph, we are left asking ourselves: What conceivable purpose *is* served by recording these patriarchal ages?

POSITIVE EVIDENCE

On the positive side the following may be urged:

1. Almost all Jewish and Christian scholars before Darwin took these chapters to be strict chronology, because they give every appearance of being such. As Professor S. R. Driver wrote: "There is a systematic chronology running through the book from the beginning almost to the end, so methodically and carefully constructed that every important birth, marriage and death has its assigned place in it."[1] Attempts are often made to drive a wedge between Ussher and the Bible, as though he said one thing and the Word of God another. But this argument would be equally valid if applied to the Athanasian Creed: the doctrine of the Trinity is nowhere stated in Scripture, but it is *logically deducible* from it. Anstey points out that "no chronologer who has adopted the numbers in the Hebrew text has ever failed to fix the Flood 1656 years, and the death of Joseph 2369 years, after Adam's creation,"[2] because these figures, and no others, are *logically deducible* from the text. Sir Isaac Newton checked Ussher's chronology, and could find no fault with it.

2. The so-called confusion about the Septuagint and Samaritan Pentateuch is really a smoke screen. "Most critical writers in modern times . . . have decided that the numbers of the Hebrew text are the most original, and therefore the most correct, on the ground that the LXX and Samaritan texts betray systematic alterations."[3] But whichever figures we accept, the indisputable fact remains

113

that *both* Greek-speaking *and* Hebrew-speaking Jews believed this to be a chronology!

3. One very interesting fact is that, in spite of all the overlapping of the patriarchs, there was *no* overlapping at the Flood. Lamech, Noah's father, died five years before it came; and Methuselah, Noah's grandfather, died in the year of the Flood. On the strict-chronology interpretation, God's timing was perfect, so that neither of the old men had to go through the ordeal. Of course this does not prove that the strict-chronology interpretation is true, but it seems to point that way.

4. Long ago Calvin perceived the purpose of these genealogies; they *are* chronology, but more than chronology. They are designed to show the continuity of revealed religion. God has had His holy prophets "since the world began" (Luke 1:70, KJV), and the continuity depends on one holy prophet's life overlapping his successor's. But according to the gap theory, Enoch could no more have talked to Methuselah than Boadicea to Queen Victoria.

COMMENTARY ON THE COMMENTATORS
No intellectual discoveries are more painful than those which expose the pedigree of ideas.

DR. L. CARMICHAEL[4]

It is instructive (though melancholy) to observe how Bible expositors have slowly but steadily been driven back from the simple and obvious interpretation of Genesis 1-11:

1. In 1824 Thomas Scott's commentary (8th ed.) was published with a full chronology accepting Ussher's dates, creation in six days, and the universal Flood.

114

2. In 1842 Dr. Gaussen of Geneva wrote: "There is no physical error in the Word of God. If there were, the book would not be from God."

3. Dr. Angus' Bible handbook (see App. C) treats Genesis 5 and 11 as strict chronology.

4. In 1865 Jamieson, Fausset, and Brown published their commentary, still accepting Ussher (at least in Genesis), but following Dr. Chalmers' "Gap Theory" between the first two verses of Genesis, in order to fit in the fossils *before* Adam; and following Dr. Pye Smith's "fantastic" *(The Genesis Flood)* theory of a local flood in Mesopotamia.

5. Patrick Fairbairn's *Imperial Dictionary of the Bible* accepts the Hebrew figures of Genesis 5 and 11 as a trustworthy chronology. The author of the article on "Chronology" writes: "The very fact that the Septuagint synchronizes with the Egyptian chronology is the strongest possible testimony against the scheme followed by the Septuagint—considering the uncertain sources whence the Egyptian chronology was deduced, the principles on which it was constructed, and the disposition so strong in that people and other ancient nations of assigning a high and even fabulous date to their origin—witness the dynasties of Manetho."[5]

6. By 1897 the archaeologists were quite emancipated from the Bible "straitjacket" and were helping themselves to thousands of (imaginary) years. So we find in Ellicott's commentary: "Scholars have *long* acknowledged that these genealogies were never intended for chronological purposes."[6]

7. Murray's *Bible dictionary* gives up all pretense of claiming the authority of Christian scholars, or biblical evidence, to deny the strict-chronology interpretation. It

roundly states that "*scientific* evidence demands a greater age . . . civilization dates back a longer period than that allowed by Ussher before Abraham."[7]

8. Dr. Richardson fairly represents the mid-twentieth century view: "The ages of the patriarchs mentioned here are, of course, entirely fanciful."[8] "These genealogical tables possess no historical value."[9]

9. Kidner accepts the life-spans of the patriarchs as literal, but rejects the chronology: "Our present knowledge of civilization, e.g. at Jericho, goes back to at least 7000 B.C., and of man himself very much further."[10] Kidner does not add, what is nevertheless a fact, that this "knowledge" is claimed by archaeologists who dismiss the Flood as a local inundation so small that it has left *no evidence whatever!* He goes on to suggest that "further study of the conventions of ancient genealogy may throw new light on the intention of the chapter."

INFALLIBLE BUT NOT INTELLIGIBLE?

This raises an important question: Is there any value in professing our belief in the infallibility of Scripture if at the same time we deny its intelligibility? Does not this border on superstition? No doubt there are some obscure verses which have baffled all the commentators; but when we find fifty-three verses of perfectly plain statements and numbers, dare we say that these words are "infallible" if they are at the same time (to us) meaningless?

Many Hindus today will swear by the divine inspiration of their Vedas; and if you point out to them the absurdity of a world supported by an elephant standing on a tortoise, they will either "spiritualize" the text or profess agnosticism on that particular point. Are we to descend to the same level of apologetic for the Bible and profess a "rever-

116

ent agnosticism'' about the years in Genesis 5 and 11, hoping that some new discovery will show that the figures do not mean what all God's people before Darwin believed them to mean? Surely the more excellent way is to challenge the *human* testimony (king lists, etc.) and chemical dating methods which appear to contradict the divine testimony.

Conclusion

The "assured results" of archaeology, where they contradict the chronology of the Bible, are no more to be trusted than the "assured results" of geology. We do not maintain that belief in the date of creation as 4004 B.C. is a necessary article of faith, but we do repudiate all the attempts of modern "science falsely so called" to reduce the figures of Genesis 5 and 11 to inexplicable doodling. Every scheme to harmonize more or less of the Bible with more or less of science eventually produces a result more or less absurd. Uniformitarian astronomers demand billions of years for starlight to reach us, uniformitarian geologists require millions of years for rocks to form, uniformitarian biologists assume millions of years for evolution, and uniformitarian archaeologists postulate thousands of years for languages to develop and buildings to fall down. None of them can prove their dates except on the basis of uniformitarianism or documents incomparably less trustworthy than the Bible. None of them can show satisfactory reasons why we should not believe what our Christian forefathers believed:

1. That the Bible is true historically, chronologically, and scientifically, wherever it touches on matters of history, chronology, or science.

2. That Adam and Eve, from whom we are all de-

scended, were created on the sixth day, about 6,000 years ago.

3. That God wiped out the "old world" by a universal Flood, between 4,000 and 5,000 years ago.

Thy word is true from the beginning.

Psalm 119:160 (KJV)

Appendix E

THE CREATION RESEARCH SOCIETY

History: Formed in 1963 as a committee of ten scientists in Michigan, it now includes more than 400 full members.

Membership: Limited to scientists having at least a graduate degree in a natural or applied science.

Statement of Belief:

1. The Bible is the written Word of God, and all its assertions are historically and scientifically true.
2. All basic types of living things were made by direct acts of God during the creation week. Whatever biological changes have occurred since then have been only within the original created kinds.
3. The great Flood described in Genesis was an historic worldwide event.

C.R.S. Quarterly: obtainable from the secretary:
Wilbert H. Rusch, Sr.
2717 Cranbrook Road
Ann Arbor, Michigan 48104

Appendix F

TIME, LIFE, AND HISTORY IN THE LIGHT OF 15,000 RADIOCARBON DATES

In the *Creation Research Quarterly* of June 1970 Robert L. Whitelaw, Professor of Nuclear and Mechanical Engineering at Virginia Polytechnic Institute, writes as follows:

"In the twenty years since introduction of radiocarbon dating by Libby, some 91 universities and laboratories in 25 different countries have dated over 15,000 independent specimens of once-living matter. Almost every imaginable form of life both recent and ancient is represented, gathered from every corner of the globe, including "prehistoric" man, a wide range of fossil flora and fauna, and even coal, petroleum and natural gas. . . . What have been the results? In a word, astounding! Astounding to every investigator with evolutionary presuppositions. But even more astounding when compared with the Biblical record—as we shall see."

A twelve-page completely documented article follows, concluding,

"1) Radiocarbon supports the idea of Biblical Creation by pointing unmistakably to a recent beginning of cosmic radiation.

"2) Radiocarbon supports a date of creation at approximately 5000 B.C.

"3) Radiocarbon supports the contemporaneous appearance of all living forms of matter at creation. Man and modern animals, along with extinct flora and fauna, all appear equally ancient and with equal suddenness.

"4) Radiocarbon supports the beginning of the human race from a few ancestors in the vicinity of the Near East.

"5) On the other hand Radiocarbon indicates the sudden concurrent appearance of the rest of the animal kingdom in larger numbers in every part of the world.

"6) Radiocarbon clearly indicates an original world in which both trees and low-lying vegetation were profuse and widespread even throughout present polar regions and deserts. [Geology and palaeontology attest an ancient world very different in climate, in location and elevation of the continents, and possibly even in the inclination of the earth's axis!]

"7) Radiocarbon points to some drastic change, shortly after creation, which depleted both animal world and arboreal vegetation, but without noticeable effect upon the multiplication of man: just such an effect as might be deduced from Genesis 3.

"8) Radiocarbon clearly points to a world-wide catastrophe destructive of man, beast and tree, just as described in Genesis 7.

"9) Radiocarbon supports the date of such a catastrophe at about 3000 B.C.

"10) Radiocarbon indicates a large and widespread human population in the world just before this catastrophe.

"11) Radiocarbon indicates the widespread existence of now-extinct flora and fauna in the world before this catastrophe, including evidence of the gradual extinction of many forms during the 2000 years between it and creation.

"12) Radiocarbon indicates that the 're-origin' of both animals and man after this catastrophe was in the vicinity of the Near East; they appeared noticeably later in the western hemisphere.

"13) Radiocarbon supports the Bible chronology of ancient empires and of Israel, and exposes the suspected exaggerations of Manetho, Berosus and others.

"14) Finally, there is no question as to which concept of Time and History is supported by the radiocarbon record. Is it the endless time and meaningless history postulated by Evolution? Or is it a specific span of time marked off by the purposeful acts of a sovereign God? . . . Fifteen thousand radiocarbon dates, dead voices from the past, assembled by scientists from every kind of once-living matter and every corner of the globe, now answer the question unequivocally in favor of the Bible!"

NOTES

CHAPTER 2

1. Henry M. Morris and John C. Whitcomb, *The Genesis Flood.*
2. Alan Dale's *Winding Quest* omits v. 11 altogether in his paraphrase of the Mosaic covenant.
3. Alan Richardson, *Genesis 1-11,* p. 50.
4. E. J. Young, *Studies in Genesis One,* pp. 55 ff.
5. Victor Pearce, *The Origin of Man.*
6. E. J. Young and F. F. Bruce, *Transactions of the Victoria Institute,* p. 21.

CHAPTER 4

1. E. J. Young, *Studies in Genesis One,* p. 95.
2. *Encyclopaedia Britannica,* 1971, s.v. "Geology."
3. Derek Kidner, *Genesis,* p. 55.
4. Henry M. Morris and John C. Whitcomb, *The Genesis Flood,* p. 136.
5. H. Enoch, *Evolution or Creation,* p. 27.
6. Fred Hoyle, *The Nature of the Universe,* pp. 35, 95, 100, 103.

CHAPTER 6

1. Fred Hoyle, *Harper's,* Apr. 1951.
2. Quoted by John C. Whitcomb in *Origin of the Solar System,* p. 22.
3. Harlow Shapley, *Beyond the Observatory.*

CHAPTER 8

1. John Calvin, *Commentary on Genesis,* pp. 172-80.
2. Bernard Ramm, *The Christian View of Science and Scripture,* pp. 93-95.
3. Henry M. Morris and John C. Whitcomb, *The Genesis Flood,* pp. 454-73.
4. C. F. Keil as quoted in ibid., p. 465.

CHAPTER 9

1. Derek Kidner, *Genesis,* p. 28.
2. Handley Moule, *Romans,* p. 60.

CHAPTER 10

1. Derek Kidner, *Genesis,* p. 66.

CHAPTER 11

1. Henry M. Morris and John C. Whitcomb, *The Genesis Flood,* p. 87.

CHAPTER 12
1. Quoted in A. Koestler, *The Sleepwalkers*, p. 535.
2. A. E. Wilder Smith, *Man's Origin, Man's Destiny*, pp. 95-103.
3. R. W. Fairbridge as quoted in Henry M. Morris and John C. Whitcomb, *The Genesis Flood*, p. 131. Itals. added.
4. Ibid., p. 153.
5. Ibid., p. 165.
6. Ibid., p. 168.
7. Ibid., p. 209.

CHAPTER 13
1. Henry M. Morris and John C. Whitcomb, *The Genesis Flood*, p. 127.
2. *Chem Tech*, May 1972, p. 296, as quoted by Whitcomb in *The World That Perished*, p. 83.
3. Morris and Whitcomb, p. 412.
4. Ibid., p. 244.
5. Ibid., p. 253

CHAPTER 14
1. Henry M. Morris and John C. Whitcomb, *The Genesis Flood*, pp. 332-78.
2. A. E. Wilder Smith, *Man's Origin, Man's Destiny*, pp. 116-26.
3. *Science*, Apr. 19, 1963.
4. *The National Geographic Society News Bulletin*, June 1961. Itals. added.
5. Morris and Whitcomb, p. 43.
6. Alexander Hislop, *The Two Babylons*, 4th ed., app. B.
7. Martin Anstey, *Romance of Bible Chronology*, p. 94.
8. Quoted in Morris and Whitcomb, p. 395.

CHAPTER 15
1. *Encyclopaedia Britannica*, 14th ed., s.v. "Language."
2. Eugene Nida, *Customs and Culture*, p. 205.
3. Richard Chevenix Trench, *On the Study of Words*, pp. 12-16.
4. *Encyclopaedia Britannica*, itals. added.
5. Nida, p. 103.

CHAPTER 16
1. F. F. Bruce, *The Spreading Flame*, Itals. added.
2. Ibid.
3. *The Sea*, Life Nature Library.

CHAPTER 17
1. Sir Gavin de Beer, *Adaptation*.
2. *Encyclopaedia Britannica*, 9th ed.

CHAPTER 18
1. Winston Churchill, *Thoughts and Adventures*, pp. 293-94.

APPENDIX B

1. Walter E. Lammerts, "Discoveries Since 1859 Which Invalidate the Evolution Theory," *Creation Research Society 1964 Annual.* Used by permission. Lammerts is Director of Research, Germain's Horticultural Research Division, Livermore, Calif.
2. Joseph Angus, *The Bible Handbook,* p. 121.

APPENDIX C

1. C. S. Lewis, *Christian Reflections.*
2. On this point the testimony of Josephus is conclusive. See Martin Anstey, *Romance of Bible Chronology,* p. 81.

APPENDIX D

1. S. R. Driver, *The Book of Genesis.*
2. Martin Anstey, *Romance of Bible Chronology,* p. 67.
3. Robert Jamieson, A. R. Fausset, David Brown, *A Commentary, Critical and Exploratory on the Old and New Testaments,* p. 84.
4. As quoted by Henry M. Morris and John C. Whitcomb, *The Genesis Flood,* p. 329. Carmichael was secretary of the Smithsonian Institute in 1953.
5. *Imperial Dictionary of the Bible,* ed. Patrick Fairbairn, s.v. "Chronology."
6. Charles J. Ellicott, *Ellicott's Commentary on the Whole Bible,* p. 55. Itals. added.
7. *Bible Dictionary.*
8. Alan Richardson, *Genesis 1-11,* p. 90.
9. Ibid., p. 123.
10. Derek Kidner, *Genesis,* p. 82.

BIBLIOGRAPHY

Allis, O. T. *God Spake by Moses*. Philadelphia: Presby. & Reformed, 1951.

Angus, Joseph. *The Bible Handbook*. New York: Revell, 1907.

Anstey, Martin. *The Romance of Bible Chronology*. London: Marshall, 1913.

Bruce, F. F. *The Spreading Flame*. London: Paternoster, 1958.

Calvin, John. *Commentaries on the First Book of Moses Called Genesis*. Grand Rapids: Eerdmans, 1948.

Churchill, Winston. *Thoughts and Adventures*.

Cousins, Frank W. *Fossil Man*. Hayling Island, Hants, Eng.: Evolution Protest Movement, 1971.

Dale, Alan. *Winding Quest*. London: Oxford U., 1974.

De Beer, Sir Gavin. *Adaptation*. Oxford Biology Reader No. 22. London: Oxford U., 1972.

Delitzsch, Franz, and Keil, Johann. *Biblical Commentary on the Old Testament*. Grand Rapids: Eerdmans, 1963.

Douglas, J. D., ed. *The New Bible Dictionary*. London: Inter-Varsity, 1962.

Driver, S. R. *The Book of Genesis*. Westminster Commentaries. London: Methuen, 1904.

Ellicott, Charles J. *Ellicott's Commentary on the Whole Bible*. Grand Rapids: Zondervan, 1954.

Encyclopaedia Britannica. 9th ed. and 14th ed.

Enoch, H. *Evolution or Creation?* London: Evangelical Press, 1972.

Fairbairn, Patrick, ed. *Imperial Dictionary of the Bible*. London: Blackie & Son, 1885. Reprint, *Fairbairn's Imperial Standard Bible Encyclopedia*. Grand Rapids: Zondervan, 1957.

Hislop, Alexander. *The Two Babylons*. 4th ed. Neptune, N.J.: Loizeaux, 1953.

Hoyle, Fred. *The Nature of the Universe*. Rev. ed. New York: Harper & Row, 1960.

Jamieson, Robert; Fausset, A. R.; and Brown, David. *A Commentary, Critical and Exploratory on the Old and New Testaments*. Rev. ed. Grand Rapids: Zondervan, 1967.

Kidner, Derek. *Genesis*. Tyndale Old Testament Commentary. Chicago: Inter-Varsity, 1967.

Koestler, A. *The Sleepwalkers*. London: Penguin Books, 1959.

Lammerts, Walter E. "Discoveries Since 1859 Which Invalidate the Evolution Theory," *Creation Research Society 1964 Annual*.

Leupold, H. C. *Exposition of Genesis*. London: Evangelical Press, 1972.

Lewis, C. S. *Christian Reflections*. Grand Rapids: Eerdmans, 1967.

————. Miracles. New York: Macmillan, 1947.

————. *Screwtape Letters*. New York: Macmillan, 1943.

Masters, P. M. *Men of Destiny*. London, Evangelical Times, 1968.

Morris, Henry M. and Whitcomb, John C. *The Genesis Flood.* London: Evangelical Press, 1961.

Moule, Handley. *Romans*. The Cambridge Bible. Cambridge: U. Press, 1879.

Murray, John, ed. *Bible Dictionary*. London: John Murray, 1908.

The National Geographic Society News Bulletin. June 1961.

Nida, Eugene. *Customs and Culture*. New York: Harper, 1954.

Ozanne, C. G. *The First 7000 Years*. Jericho, N.Y.: Exposition, 1970.

Pearce, Victor. *The Origin of Man*. London: Crusade Magazine, 1967.

Ramm, Bernard. *The Christian View of Science and Scripture*. Grand Rapids: Eerdmans, 1954.

Richardson, Alan. *Genesis 1-11*. Torch Bible Commentaries. London: SCM, 1953.

Science, Apr. 19, 1963.

The Sea. Life Nature Library. Netherlands: Time-Life International, 1963.

Shapley, Harlow. *Beyond the Observatory*. New York: Scribner, 1972.

Smith, A. E. Wilder. *The Creation of Life*. Wheaton, Ill.: Harold Shaw, 1970.

_____. *Man's Origin, Man's Destiny*. Wheaton, Ill.: Harold Shaw, 1968.

Solzhenitsyn, Alexander. *Cancer Ward*. New York: Dial, 1968.

Trench, Richard Chevenix. *On the Study of Words*. 1892. Reprint. Detroit: Gale Research, n.d.

von Rad, Gerhard. *Genesis, a Commentary*. New York: Harper & Row, 1960.

Whitcomb, J. C. *The Early Earth*. London: Evangelical Press, 1972.

_____. *Origin of the Solar System*. Philadelphia: Presby. & Reformed, 1974.

_____. *The World That Perished*. London: Evangelical Press, 1973.

White, A. J. Monty. *Radio-Carbon Dating*. Glamorgan, Wales: A. J. Monty White, n.d.

Wright, G. E. *Biblical Archaeology*. Philadelphia: Westminster, 1957.

Young, E. J. *Genesis Three*. London: Banner of Truth, 1966.

_____. *Studies in Genesis One*. Philadelphia: Presby. & Reformed, 1964.

Young, E. J. and Bruce, F. F. *Transactions of the Victoria Institute*. London: Victoria Institute, 1946.

Zimmerman, Paul A. *Darwin, Evolution, and Creation*. St. Louis: Concordia, 1966.